INTRODUCTION TO GROUNDWATER MODELING

A SERIES OF BOOKS IN GEOLOGY
Editor: Allan Cox

INTRODUCTION TO GROUNDWATER MODELING

Finite Difference and Finite Element Methods

Herbert F. Wang
Mary P. Anderson

UNIVERSITY OF WISCONSIN, MADISON

W. H. FREEMAN AND COMPANY
San Francisco

Project Editor: Larry Olsen
Copy Editor: Ruth Cottrell
Designer: Robert Ishi
Production Coordinator: Bill Murdock
Illustration Coordinator: Cheryl Nufer
Artist: Georg Klatt
Compositor: Syntax International
Printer and Binder: The Maple-Vail Book Manufacturing Group

Library of Congress Cataloging in Publication Data

Wang, Herbert.
 Introduction to groundwater modeling.

 Bibliography: p.   **228113**
 Includes index.
  1. Groundwater flow—Mathematical models.
2. Difference equations. 3. Finite element method.
I. Anderson, Mary P. II. Title.
TC176.W36   551.49′0724   81-2665
ISBN 0-7167-1303-9   AACR2

Printed in the United States of America

987654321   MP   0898765432

Contents

Preface

Mathematical models of groundwater flow have been used since the late 1800s. A mathematical model consists of a set of differential equations that are known to govern the flow of groundwater. The reliability of predictions from a groundwater model depends on how well the model approximates the field situation. Inevitably, simplifying assumptions must be made in order to construct a model because the field situation is too complex to be simulated exactly. Usually, the assumptions necessary to solve a mathematical model analytically are fairly restrictive—for example, many analytical solutions require that the medium be homogeneous and isotropic. To deal with more realistic situations, it is usually necessary to solve the mathematical model approximately using numerical techniques. Since the 1960s, when high-speed digital computers became widely available, numerical models have been the favored type of model for studying groundwater. The subject of this book is the use of numerical models to simulate groundwater flow and contaminant transport.

This book offers a fundamental and practical introduction to finite difference and finite element techniques. Our goal is to enable readers to solve groundwater flow problems with the digital computer, and every topic is developed with the aim of conveying a full understanding of the steps leading to the short sample computer programs included as part of the text. The programs can be run on any computer with a FORTRAN compiler. (On the University of Wisconsin

Univac 1100, job charges were about 75¢ per program.) Several of the sample problems appear in different forms throughout the text to illustrate various methods and assumptions. Problems at the ends of chapters are designed to reinforce the principles presented in the text.

The book covers five major topics. In Chapter 1, we review some fundamental principles of groundwater flow. In Chapters 2 and 3, we present an introduction to the finite difference method as applied to steady-state problems. Our method is to present selected applications for which numerical solutions are compared with analytical solutions. This method is used to verify the accuracy of the numerical solution. Once we have established confidence in the numerical solution technique, the numerical model can be used to solve problems for which no analytical solutions are available.

In Chapters 4 and 5, the method of finite differences is applied to transient flow problems. Governing equations used in Chapters 1 through 5 are derived as needed. Chapters 6 and 7 contain an introduction to the method of finite elements as applied to steady-state and transient flow problems, respectively. Finally, Chapter 8 contains a discussion of contaminant transport. We derive the advection-dispersion equation, which governs the movement of contaminants through a groundwater system, and we use the finite element method to solve a sample problem.

The subject matter becomes progressively more difficult in the later chapters of the book, and readers should expect to spend more time comprehending material in Chapters 4 and 5 than in Chapters 1 through 3. Likewise, the material in Chapters 6 through 8 is intrinsically more difficult than that in the earlier chapters.

Our book can serve as the first introduction for readers headed towards advanced work in numerical modeling of groundwater systems, or it can serve as a complete course for readers headed towards related areas of water resources who need a basic grasp of modeling concepts. Calculus, physics, FORTRAN programming, and a brief introduction to matrices are necessary prerequisites. The book can be used in a one-semester senior or graduate level course in geology or engineering. It could also supplement more general courses in hydrogeology or fluid mechanics. Professional engineers and geologists who desire an introduction to groundwater modeling should find the book readable and useful, especially if they have access to a computer.

We are especially grateful to Irwin Remson, who critically reviewed several versions of the entire manuscript. John Bredehoeft, Jay Lehr, Debu Majumdar, and Evelyn Roeloffs also made helpful suggestions. Thanks also go to the many students who commented on early versions of the manuscript.

June 1981

Herbert F. Wang
Mary P. Anderson

INTRODUCTION TO GROUNDWATER MODELING

CHAPTER 1

Introduction

1.1 MODELS

A model is a tool designed to represent a simplified version of reality. Given this broad definition of a model, it is evident that we all use models in our everyday lives. For example, a road map is a way of representing a complex array of roads in a symbolic form so that it is possible to test various routes on the map rather than by trial and error while driving a car. A road map could be considered a kind of model (Lehr, 1979) because it is a way of representing reality in a simplified form. Similarly, groundwater models are also representations of reality and, if properly constructed, can be valuable predictive tools for management of groundwater resources. For example, using a groundwater model, it is possible to test various management schemes and to predict the effects of certain actions. Of course, the validity of the predictions will depend on how well the model approximates field conditions. Good field data are essential when using a model for predictive purposes. However, an attempt to model a system with inadequate field data can also be instructive as it may serve to identify those areas where detailed field data are critical to the success of the model. In this way, a model can help guide data collection activities.

Types of Groundwater Models

Several types of models have been used to study groundwater flow systems. They can be divided into three broad categories (Prickett, 1975): *sand tank models, analog models,* including viscous fluid models and electrical models, and *mathematical models,* including analytical and numerical models. A sand tank model consists of a tank filled with an unconsolidated porous medium through which water is induced to flow. A major drawback of sand tank models is the problem of scaling down a field situation to the dimensions of a laboratory model. Phenomena measured at the scale of a sand tank model are often different from conditions observed in the field, and conclusions drawn from such models may need to be qualified when translated to a field situation.

As we shall see later in the book, the flow of groundwater can be described by differential equations derived from basic principles of physics. Other processes, such as the flow of electrical current through a resistive medium or the flow of heat through a solid, also operate under similar physical principles. In other words, these systems are analogous to the groundwater system. The two types of analogs used most frequently in groundwater modeling are viscous fluid analog models and electrical analog models.

Viscous fluid models are known as Hele–Shaw or parallel plate models because a fluid more viscous than water (for example, oil) is made to flow between two closely spaced parallel plates, which may be oriented either vertically or horizontally. Electrical analog models were widely used in the 1950s before high-speed digital computers became available. These models consist of boards wired with electrical networks of resistors and capacitors. They work according to the principle that the flow of groundwater is analogous to the flow of electricity. This analogy is expressed in the mathematical similarity between Darcy's law for groundwater flow and Ohm's law for the flow of electricity. Changes in voltage in an electrical analog model are analogous to changes in groundwater head. A drawback of an electrical analog model is that each one is designed for a unique aquifer system. When a different aquifer is to be studied, an entirely new electrical analog model must be built.

A mathematical model consists of a set of differential equations that are known to govern the flow of groundwater. Mathematical models of groundwater flow have been in use since the late 1800s. The reliability of predictions using a groundwater model depends on how well the model approximates the field situation. Simplifying assumptions must always be made in order to construct a model because the field situations are too complicated to be simulated exactly. Usually the assumptions necessary to solve a mathematical model analytically are fairly restrictive. For example, many analytical solutions require that the medium be homogeneous and isotropic. To deal with more realistic situations, it is usually necessary to solve the mathematical model approximately using numerical techniques. Since the 1960s, when high-speed digital computers became widely available, numerical models have been the favored type of model for studying groundwater. The subject of this book is the use of numerical methods to solve mathematical models that simulate groundwater flow and contaminant transport.

We consider two types of models—finite difference models (Chapters 2 through 5) and finite element models (Chapters 6 through 8). In either case, a system of nodal points is superimposed over the problem domain. For example, consider the problem shown in Figure 1.1. The problem domain consists of an aquifer bounded on one side by a river (Figure 1.1a). The aquifer is recharged areally by precipitation, but there is no horizontal flow out of or into the aquifer except along the river. Two finite difference representations of the problem domain are illustrated in Figures 1.1b and 1.1c, and a finite element representation is shown in Figure 1.1d.

The concept of elements (that is, the subareas delineated by the lines connecting nodal points) is fundamental to the development of equations in the finite element method. Triangular elements are used in Figure 1.1d, but quadrilateral or other elements are also possible. In the finite difference method, nodes may be located inside cells (Figure 1.1b) or at the intersection of grid lines (Figure 1.1c). The finite difference grid shown in Figure 1.1b is said to use block-centered nodes, whereas the grid in Figure 1.1c is said to use mesh-centered nodes. Aquifer properties and head are assumed to be constant within each cell in Figure 1.1b. In Figure 1.1c, nodes are located at the intersections of grid lines, and the area of influence of each node is defined following one of several different conventions. Regardless of the representation, an equation is written in terms of each nodal point because the area surrounding a node is not directly involved in the development of finite difference equations.

The goal of modeling is to predict the value of the unknown variable (for example, groundwater head or concentration of a contaminant) at nodal points. Models are often used to predict the effects of pumping on groundwater levels. For example, consider the aquifer shown in Figure 1.1. In this example, a model could be used to predict the effects of pumping the three wells in the well field on water levels in the four observation wells or to predict the effects of installing additional pumping wells. The model could also be used to determine how much water would be diverted from the river as a result of pumping. However, before a predictive simulation can be made, the model should be calibrated and verified. The process of calibrating and verifying a model is discussed in Chapter 5.

4

(a)

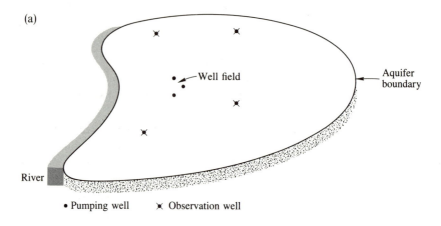

River

• Pumping well   ✕ Observation well

(b)

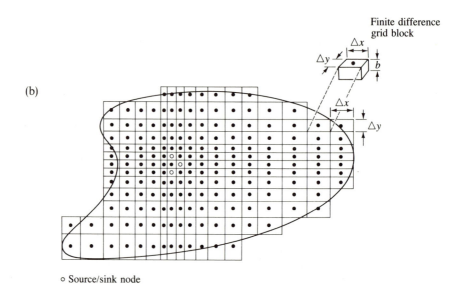

Finite difference
grid block

∘ Source/sink node

Figure 1.1
Finite difference and finite element representations of an aquifer region.
(a) Map view of aquifer showing well field, observation wells, and boundaries.
(b) Finite difference grid with block-centered nodes, where Δx is the spacing in the x direction, Δy is the spacing in the y direction, and b is the aquifer thickness.

(c)

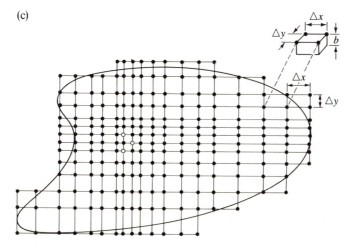

∘ Source/sink node

(d)

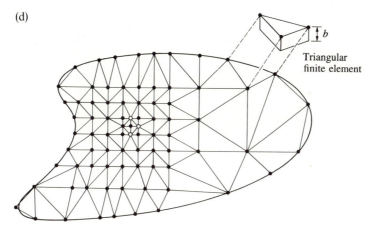

Triangular
finite element

∘ Source/sink node

(c) Finite difference grid with mesh-centered nodes.
(d) Finite element mesh with triangular elements where b is the aquifer thickness.
(Adapted from Mercer and Faust, 1980a.)

1.2 PHYSICS OF GROUNDWATER FLOW

Darcy's Law

Darcy set out to find experimentally what factors govern water flow through a sand filter (Figure 1.2). He measured the discharge by timing the rate at which water filled a 1 square meter basin at the outlet, and he measured the head drop across the sand. Darcy defined head to be the height, relative to the elevation of the bottom of the sand, to which water rises in each U-shaped tube. Although Darcy used mercury-filled manometers, he always reported his head data in terms of the equivalent water height. We shall demonstrate that head is proportional to the sum of the pressure potential of the mercury (or any fluid) in the U-shaped tube plus the elevation potential relative to the

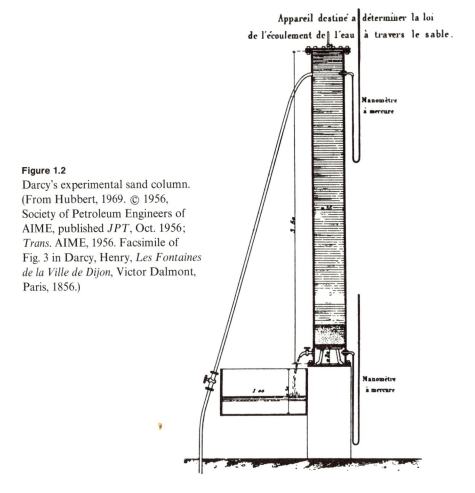

Figure 1.2
Darcy's experimental sand column.
(From Hubbert, 1969. © 1956,
Society of Petroleum Engineers of
AIME, published *JPT*, Oct. 1956;
Trans. AIME, 1956. Facsimile of
Fig. 3 in Darcy, Henry, *Les Fontaines
de la Ville de Dijon*, Victor Dalmont,
Paris, 1856.)

base level. Applying the term head to the height above sea level of water in a well is the correct field use in Darcy's original sense of the term.

By a series of experiments, Darcy established that, for a given type of sand, the volume discharge rate Q is directly proportional to the head drop $h_2 - h_1$ and to the cross-sectional area A, but it is inversely proportional to the length difference $\ell_2 - \ell_1$. Calling the proportionality constant K the hydraulic conductivity gives Darcy's law:

$$Q = -KA \frac{h_2 - h_1}{\ell_2 - \ell_1} \tag{1.1}$$

The negative sign signifies that groundwater flows in the direction of head loss. Figure 1.3 is a graph showing Darcy's experimental data. It illustrates the linear relationship between discharge rate and head drop for two different sands.

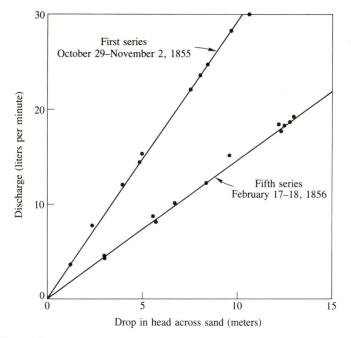

Figure 1.3
Darcy's data showing that discharge is proportional to head drop for two different sands. (From Hubbert, 1969. © 1956, Society of Petroleum Engineers of AIME, published *JPT*, Oct. 1956; *Trans.* AIME, 1956.)

Hubbert's Force Potential

Groundwater flows in response to pressure differences and elevation differences. Numerous persons made the error of equating head to pressure and neglecting elevation. Hubbert (1940) clarifies the concept of groundwater potential and its relationship to Darcy's head by deriving it from basic physical principles. Groundwater potential at a given point is the energy required to transport a unit mass of water from a standard reference state to that point. Differences in potential give rise to groundwater flow; that is, water moves from higher potential to lower potential. The potential is called a force potential because its space derivative has units of force per unit mass.

We now follow Hubbert's derivation of the groundwater potential. Two separate force potentials—pressure and elevation—act on a unit mass of groundwater. Suppose we have a sand-filled tube saturated with water, and the pressure is P at a height z. The potential energy per unit mass of water is defined to be the work required to bring a unit mass of water from a reference position z_{ref} to its actual position z. If we consider the pressure at the reference position to be zero, then the pressure P is in gage pressure, the pressure above atmospheric.

We consider separately the work required to raise the unit mass of water to pressure P and to raise the unit mass to elevation z. The work per unit mass required to raise the water pressure is

$$W = \frac{1}{m} \int_0^P V \, dP \tag{1.2}$$

where m is the water mass and V is the water volume. The volume V is m/ρ_w, where ρ_w is the density of water. If the water is assumed to be incompressible, that is, density is the same at all pressures, then the work per unit mass to raise the water pressure to P is P/ρ_w. The work per unit mass required to raise the fluid to elevation z is $g(z - z_{ref})$, where g is the acceleration of gravity. Therefore, the total groundwater potential is

$$\phi = \frac{P}{\rho_w} + g(z - z_{ref}) \tag{1.3}$$

We have expressed the sought-for potential for groundwater flow in fundamental physical terms. How is the potential ϕ related to Darcy's head h? That is, how do the terms of Equation 1.3 relate to the physically measured quantities in Darcy's experiment? Refer to Figure 1.4. Let the elevation reference datum

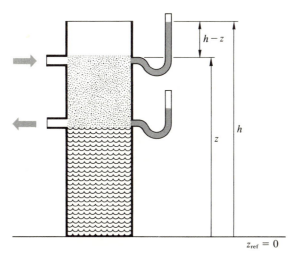

Figure 1.4
Total head (h) as the sum of pressure head ($h - z$) and elevation head (z).

be z_{ref}. Then the water pressure P at an elevation equal to z is $P = \rho_w g(h - z)$. If this expression is substituted into Equation 1.3, and if $z_{ref} = 0$, then

$$\phi = gh \tag{1.4}$$

Keep in mind that ϕ and h are functions of elevation z. Equation 1.4 tells us that the potential ϕ derived from basic fluid mechanics is directly proportional to Darcy's experimental head. In fact, head can be thought of as a potential expressed in terms of energy per unit weight of water, whereas Hubbert's potential ϕ is expressed in terms of energy per unit mass. We can also think of the separate terms, pressure head and elevation head, as having units of energy per unit weight; that is,

$$h = \frac{P}{\rho_w g} + z \tag{1.5}$$

where h is total head, $P/\rho_w g$ is pressure head, and z is elevation head. The head drop in Darcy's law is proportional to the energy loss that results from friction of fluid flowing against the sides of the pore channels. Darcy's law is an expression of the fact that groundwater moves in a direction of decreasing energy or from higher to lower head.

Darcy's Law in Three Dimensions

We now want to generalize the potential to be a function of all three space coordinates, that is, $h = h(x, y, z)$, and to generalize $dh/d\ell$, the rate of change of head with position, to three dimensions. So far we have considered only the vertical direction, where $h = h(z)$. We assume an isotropic porous medium and assume that the discharge rate Q is not dependent on time. In Appendix A, Darcy's law in two dimensions is generalized to an anisotropic medium.

Let us define $q = Q/A$ to be the volume rate of flow per unit area. The quantity q is called the specific discharge. In the limit, as the head drop $h_2 - h_1$ occurs over a smaller and smaller interval $\ell_2 - \ell_1$, we can write Darcy's law in the differential form

$$q = -K \frac{dh}{d\ell} \tag{1.6}$$

The specific discharge q has units of velocity and is known also as the Darcy velocity. The average linear or pore velocity v is q/n, where n is porosity.

The three-dimensional generalization of Darcy's law requires that the one-dimensional form, Equation 1.6, be true for each of the x, y, and z components of flow:

$$q_x = -K \frac{\partial h}{\partial x} \qquad q_y = -K \frac{\partial h}{\partial y} \qquad q_z = -K \frac{\partial h}{\partial z} \tag{1.7}$$

Note that the space derivatives are partial derivatives because head is now a function of all three space coordinates. Equation 1.7 can be written in the shorthand of vector notation as

$$\mathbf{q} = -K \, \mathbf{grad} \, h \tag{1.8}$$

The velocity vector $\mathbf{q}$ has components q_x, q_y, and q_z, and the gradient vector $\mathbf{grad} \, h$ has components $\partial h/\partial x$, $\partial h/\partial y$, and $\partial h/\partial z$. Because each component of $\mathbf{q}$ is the same scalar multiple K of the corresponding component of $(-\mathbf{grad} \, h)$, the vectors $\mathbf{q}$ and $(-\mathbf{grad} \, h)$ both point in the same direction. This conclusion follows from the assumption of isotropy. For an arbitrary direction in an anisotropic medium, the velocity vector $\mathbf{q}$ does not, in general, point in the same direction as $(-\mathbf{grad} \, h)$ (see Appendix A).

Let us consider a two-dimensional situation $h = h(x, y)$ to illustrate properties of the gradient of head. Contour lines of head are defined by the condition

$$\Delta h = \frac{\partial h}{\partial x} \Delta x + \frac{\partial h}{\partial y} \Delta y = 0 \tag{1.9}$$

The slope of the tangent $\Delta y / \Delta x$ to a contour line is equal to $-(\partial h/\partial x)/(\partial h/\partial y)$. The slope of **grad** h, on the other hand, is equal to the ratio of its components $(\partial h/\partial y)/(\partial h/\partial x)$. The slope of the gradient vector times the direction of the contour line at each point is negative 1. Therefore, the gradient vector is perpendicular to the contour of constant head. The flow is perpendicular to the contour lines because, for an isotropic medium, the velocity vector is in the direction of the negative of the gradient vector.

Continuity Equation for Steady-State Flow

Darcy's law, Equation 1.8, summarizes much of the physics of groundwater flow by relating the velocity vector to the gradient of potential. Continuity or conservation is a second important law. For steady-state conditions, continuity requires that the amount of water flowing into a representative elemental volume be equal to the amount flowing out. The existence of steady-state conditions implies that head is independent of time. For now, we make two other simplifying assumptions. First, we assume water is incompressible. If it were not, it could be compressed and stored in the elemental volume. Mass is always conserved, but the volume is conserved only for an incompressible liquid. Second, we assume the elemental volume contains no sources or sinks. This means we will not allow water to be added to or removed from the aquifer by means of wells or by precipitation or transpiration, for example.

We begin the analysis by considering the flow into and out of an elemental cube whose sides are of length Δx, Δy, and Δz (Figure 1.5). The volume of the cube is $\Delta V = \Delta x \, \Delta y \, \Delta z$. The mass balance is computed by summing the results from each component direction. For example, the component q_y represents the volume rate of flow per unit area through the left face of the cube. The discharge through the left and right faces is the product of the flow rate per unit area times the area, $\Delta x \, \Delta z$, of the face through which the flow occurs. At the right face, q_y is different by an amount $(\partial q_y/\partial y) \Delta y$. Therefore, the net change in the discharge rate in the y direction is $(\partial q_y/\partial y) \Delta y (\Delta x \, \Delta z) = (\partial q_y/\partial y) \Delta V$. The sign convention is that if the derivative is positive, there is a net outward flow of water. We can apply a similar analysis for the x and z directions to obtain

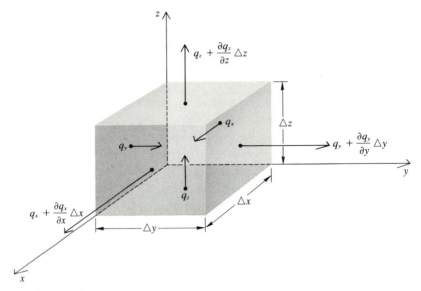

Figure 1.5
Analysis of net volume flow per unit time through an infinitesimal cube.

the net change in the discharge rate in the x direction to be $(\partial q_x/\partial x)\,\Delta V$ and the net change in the discharge in the z direction to be $(\partial q_z/\partial z)\,\Delta V$. The sum, $(\partial q_x/\partial x)\,\Delta V + (\partial q_y/\partial y)\,\Delta V + (\partial q_z/\partial z)\,\Delta V$, must equal zero. We can divide through by ΔV and be left with the continuity equation for steady-state conditions

$$\frac{\partial q_x}{\partial x} + \frac{\partial q_y}{\partial y} + \frac{\partial q_z}{\partial z} = 0 \qquad (1.10)$$

The left-hand side of Equation 1.10 represents the net change in the volume rate of flow per unit volume. As such, it is called the divergence of $\mathbf{q}$ and written

$$\operatorname{div} \mathbf{q} = \frac{\partial q_x}{\partial x} + \frac{\partial q_y}{\partial y} + \frac{\partial q_z}{\partial z} \qquad (1.11)$$

Divergence is an operation performed on a vector that produces a scalar quantity. Gradient, on the other hand, is an operation performed on a scalar that produces a vector quantity.

1.3 LAPLACE'S EQUATION

Laplace's equation combines Darcy's law and the continuity equation into a single second-order partial differential equation. Darcy's law is substituted component by component into Equation 1.10 to give

$$\frac{\partial}{\partial x}\left(-K\frac{\partial h}{\partial x}\right) + \frac{\partial}{\partial y}\left(-K\frac{\partial h}{\partial y}\right) + \frac{\partial}{\partial z}\left(-K\frac{\partial h}{\partial z}\right) = 0 \qquad (1.12)$$

where $K = K(x, y, z)$. If K is assumed to be independent of x, y, and z—that is, if the region is assumed to be homogeneous as well as isotropic—then Equation 1.12 becomes

$$\frac{\partial^2 h}{\partial x^2} + \frac{\partial^2 h}{\partial y^2} + \frac{\partial^2 h}{\partial z^2} = 0 \qquad (1.13)$$

Equation 1.13 is Laplace's equation—the governing equation for groundwater flow through an isotropic, homogeneous aquifer under steady-state conditions. Laplace's equation is also used in other branches of physics. For example, it is the governing equation for heat conduction in a solid under steady-state conditions.

Boundary Conditions

What does it mean to solve Laplace's equation? The physical quantity of interest is the head as a function of x, y, and z. Laplace's equation states only that the sum of the second partial derivatives of h with respect to x, y, and z is zero. The solution of Laplace's equation requires specification of boundary conditions which constrain the problem and make solutions unique. The different types of boundary conditions are: (a) head is known for surfaces bounding the flow region (Dirichlet conditions); (b) flow is known across surfaces bounding the region (Neumann conditions); (c) some combination of (a) and (b) is known for surfaces bounding the region (mixed conditions). The groundwater hydrologist must sometimes approximate boundary conditions to limit the region of the problem domain. If inconsistent or incomplete boundary conditions are specified, the problem itself is ill defined.

1.4 REGIONAL GROUNDWATER FLOW SYSTEM

As an example of a groundwater model involving Laplace's equation and suitable boundary conditions, we present a regional groundwater problem described by Toth (1962). He was able to draw conclusions about the configuration of regional groundwater systems by using a mathematical model.

Figure 1.6 represents a cross section through a small watershed bounded on one side by a topographic high, which marks a regional groundwater divide, and on the other side by a major stream, which is a groundwater discharge area and marks another regional groundwater divide. The aquifer is assumed to consist of homogeneous, isotropic, porous material underlain by impermeable rock.

We first consider the boundary conditions. The left and right groundwater divides can be represented mathematically as impermeable, no-flow boundaries. Although no physical barrier exists, a groundwater divide has the same effect as an impermeable barrier because no groundwater crosses it. Groundwater to the right of the valley bottom discharges at point *A*, and groundwater on either side of the topographic high flows away from point *B*. The lower boundary is also a no-flow boundary because the impermeable basement rock forms a physical barrier to flow. The upper boundary of the mathematical model is

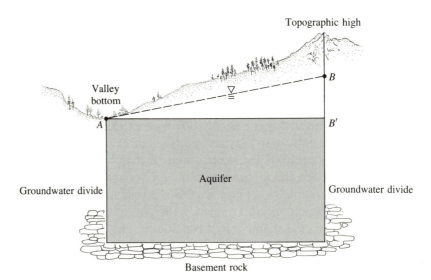

Figure 1.6
Schematic representation of the boundaries of a two-dimensional regional groundwater flow system.

the horizontal line AB' even though the water table of the physical system lies above AB'. Thus the rectangular problem domain of the mathematical model is an approximation to the actual shape of the saturated flow region. Along the boundary AB', the head is taken to be equal to the height of the water table, and the water table configuration is considered to be a straight line.

Toth (1962, 1963) finds that this mathematical model is a realistic representation of the general configuration of the flow system where the topography is subdued and the water table slope is gentle. Toth (1963) also uses a more general expression for the configuration of the water table in a region of gently rolling topography.

We must express the boundary conditions shown in Figure 1.6 in mathematical terms. The coordinate system is defined in Figure 1.7. An equation is required for each boundary. Consider the upper boundary first. The boundary is located at $y = y_0$ for x ranging from 0 to s. The distribution of head along this boundary is assumed to be linear. The equation for a linear variation such that $h(0, y_0) = y_0$ is $h(x, y_0) = cx + y_0$ for $0 \le x \le s$, where c is the slope of the water table. The specification of head along the upper boundary makes it a Dirichlet boundary condition.

The other three boundary conditions are for no-flow boundaries. Darcy's law relates flow to gradient of head. Along a vertical, no-flow boundary, $q_x = 0$ implies $\partial h / \partial x = 0$, and along a horizontal, no-flow boundary, $q_y = 0$

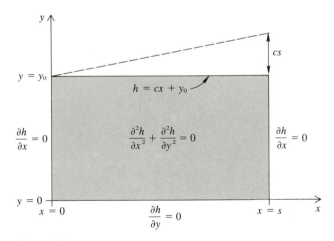

Figure 1.7
Mathematical model of the regional groundwater flow system shown in Figure 1.6.

implies $\partial h/\partial y = 0$. Specification of flow across these three boundaries makes them Neumann boundary conditions. In this example, we have specified both head and flow conditions, and, therefore, we have a mixed problem. The full set of boundary conditions is written as follows.

Top $\qquad$ $h(x, y_0) = cx + y_0$ $\qquad$ $0 \leq x \leq s$

Bottom $\qquad$ $\left.\dfrac{\partial h}{\partial y}\right|_{y=0} = 0$ $\qquad$ $0 \leq x \leq s$

Left $\qquad$ $\left.\dfrac{\partial h}{\partial x}\right|_{x=0} = 0$ $\qquad$ $0 \leq y \leq y_0$

Right $\qquad$ $\left.\dfrac{\partial h}{\partial x}\right|_{x=s} = 0$ $\qquad$ $0 \leq y \leq y_0$

Now we consider the governing equation. We know that Laplace's equation simulates groundwater flow in a homogeneous, isotropic aquifer if there is no accumulation or loss of water within the system. Toth assumed that, in an undeveloped watershed, the fluctuations of the water table on an annual basis were small. That is, he used an average water table position and assumed that the system was at a steady state on an annual basis. The water table position at the beginning of the year was the same as the position at the end of the year; there was no net accumulation or loss of water from the system. Therefore, with this idealization, the two-dimensional Laplace equation, $\partial^2 h/\partial x^2 + \partial^2 h/\partial y^2 = 0$, is the required governing equation. The mathematical model, consisting of the governing equation together with the four boundary conditions, is summarized in Figure 1.7. The solution of this model is presented in Chapter 2.

Notes and Additional Reading

1. A literature review describing the use of groundwater flow models may be found in Prickett (1975). Viscous fluid and electrical analog models, as well as mathematical models, are described in detail. For an interesting application of a vertically oriented parallel plate model, see Collins et al. (1972). Rushton and Redshaw (1979) discuss electrical analog models in detail, and Getzen (1977) describes an application of an electrical analog model to the analysis of three-dimensional groundwater flow in Long Island.

2. A general overview of the concept of a model as applied to groundwater is contained in Mercer and Faust (1980a), and various governing equations are considered by Mercer and Faust (1980b). Faust and Mercer (1980a) discuss the distinction between the method of finite differences and finite elements. These papers and others (Mercer and Faust, 1980c; Faust and Mercer, 1980b) have been collected into Mercer and Faust (1981).

3. Fundamental groundwater physics is presented at an elementary level by Bennett (1976). The text is presented in a workbook format with questions and answers. Parts I, II, and III pertain to material covered in our Chapter 1. Part I covers definitions and general concepts. Part II covers Darcy's law, and Part III covers applications of Darcy's law to field problems.

4. Remson et al. (1971) is an intermediate-to-advanced-level book on the use of numerical methods with emphasis on finite difference techniques. In Chapter 1 of Remson et al. (pp. 6–19), various types of potentials are discussed, and Hubbert's force potential is derived.

5. Most standard texts in groundwater hydrology include a discussion of the physics of groundwater flow. For a good discussion of Darcy's law and the concept of potential, see Freeze and Cherry (1979, pp. 15–38) and Fetter (1980, pp. 106–117). Freeze and Cherry (pp. 63–64) present a slightly different derivation of Laplace's equation.

6. Groundwater flow, electrical current flow, and heat flow are transport processes that occur in a resisting medium. A summary of the electrical and heat flow analogies is given in Appendix D, Table D.1. In a mathematical sense, the description of each problem is identical.

CHAPTER 2

Finite Differences: Steady-State Flow (Laplace's Equation)

2.1 INTRODUCTION

A mathematical groundwater model for steady-state conditions consists of a governing equation and boundary conditions which simulate the flow of groundwater in a particular problem domain. To solve the model requires calculating the values of head at each point in the system. Using techniques of calculus, it is sometimes possible to write an expression for head as a function of the space coordinates. Then we have an analytical solution. For example, Toth (1962) presents an analytical solution to the regional groundwater flow system model discussed in Chapter 1 (Figure 1.7):

$$h(x, y) = y_0 + \frac{cs}{2} - \frac{4cs}{\pi^2} \sum_{m=0}^{\infty} \frac{\cos[(2m + 1)\pi x/s] \cosh[(2m + 1)\pi y/s]}{(2m + 1)^2 \cosh[(2m + 1)\pi y_0/s]} \quad (2.1)$$

To obtain this analytical solution, Toth (1962) introduces several simplifying assumptions. For example, he assumes a homogeneous and isotropic aquifer plus a linear water table configuration, and he approximates the problem domain by a rectangle. None of these assumptions is necessary to obtain a numerical solution.

For many problems, the assumptions that must be made to obtain an analytical solution will not be realistic. In these cases, we must resort to approximate methods using numerical techniques to solve the mathematical model. The two numerical techniques that we consider are the method of finite differences and the method of finite elements. These two approximate methods provide a rationale for operating on the differential equations that make up a model and for transforming them into a set of algebraic equations. Before digital computers were widely available, only hand calculations were possible, and approximating techniques were of limited value. Using computers, one can solve large numbers of algebraic equations by iterative techniques or by direct matrix methods. We can test a numerical solution by comparing the head distribution generated by the computer with that determined from an analytical solution, if one is available, and we can also check the simulated head distribution with the values observed in the field. The procedure is summarized in Figure 2.1.

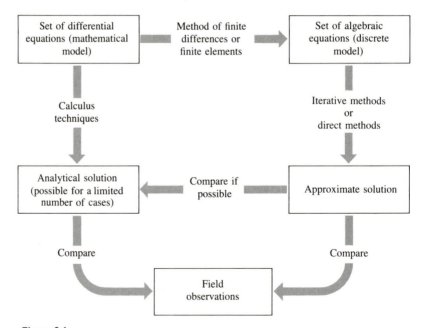

Figure 2.1
Relationships between mathematical model, discrete algebraic model, analytical solution, approximate solution, and field observations.

2.2   DIFFERENCES FOR DERIVATIVES

Finite Difference Expression of Laplace's Equation

Analytical solutions can be used to calculate values for the unknown at any point in the problem domain. Numerical solutions yield values for only a predetermined, finite number of points in the problem domain. By limiting our need to know the head to a reasonable number of points N, we can convert a partial differential equation into a set of N algebraic equations involving N unknown potentials. In this section, we derive the finite difference approximation to Laplace's equation in two dimensions.

Consider a finite set of points on a regularly spaced grid (Figure 2.2). Lattice points are spaced horizontally by a distance Δx and vertically by a distance Δy. The distances Δx and Δy are the natural units of the lattice. The smaller we make Δx and Δy, the closer the approximate solution comes to the analytical solution. The trade-off is that the number of unknowns increases. To locate any point in the grid, we specify an integer ordered pair (i, j). Relative to the ordered pair $(1, 1)$ located in the upper left-hand corner, (i, j) is a distance $(i - 1)\Delta x$ in the positive x direction and $(j - 1)\Delta y$ in the negative y direction. This

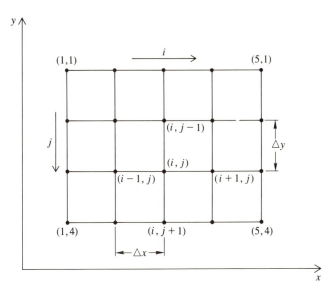

Figure 2.2
Finite difference grid showing index numbering convention.

indexing convention inverts rows and columns when compared with standard matrix notation. Readers should always be alert to different conventions for the meaning of the ordered pair (i, j).

The value of the head at the point represented by the indices (i, j) is $h_{i,j}$. Sometimes we omit the comma between the subscripts if no ambiguity results. Let the Cartesian coordinates (x_0, y_0) be represented by (i, j). Along the horizontal line $y = y_0$, consider a profile of head which has in succession the values $h_{i-1,j}$, $h_{i,j}$, and $h_{i+1,j}$. In the finite difference approximation, derivatives are replaced by differences taken between nodal points. A central approximation to $\partial^2 h / \partial x^2$ at (x_0, y_0) is obtained by approximating the first derivative at $(x_0 + \Delta x/2, y_0)$ and at $(x_0 - \Delta x/2, y_0)$, and then obtaining the second derivative by taking a difference between the first derivatives at those points. That is,

$$\frac{\partial^2 h}{\partial x^2} \simeq \frac{\dfrac{h_{i+1,j} - h_{i,j}}{\Delta x} - \dfrac{h_{i,j} - h_{i-1,j}}{\Delta x}}{\Delta x} \tag{2.2}$$

which simplifies to

$$\frac{\partial^2 h}{\partial x^2} \simeq \frac{h_{i-1,j} - 2h_{i,j} + h_{i+1,j}}{(\Delta x)^2} \tag{2.3}$$

Similarly,

$$\frac{\partial^2 h}{\partial y^2} \simeq \frac{h_{i,j-1} - 2h_{i,j} + h_{i,j+1}}{(\Delta y)^2} \tag{2.4}$$

According to Laplace's equation, we must add the preceding two equations and set the result equal to zero. If we consider a square grid of points such that $\Delta x = \Delta y$, then the finite difference approximation for Laplace's equation at the point (i, j) simplifies to

$$h_{i-1,j} + h_{i+1,j} + h_{i,j-1} + h_{i,j+1} - 4h_{i,j} = 0 \tag{2.5}$$

Equation 2.5, or a generalized form of it, is the most widely used equation in finite difference solutions of steady-state flow problems, and it will form the heart of a computer program. There will be one equation of the form of Equation 2.5 for each interior point (i, j) of the problem.

Example with Dirichlet Boundary Conditions: Region near a Well

We present a Dirichlet problem to illustrate how Equation 2.5 leads to one algebraic equation for each unknown potential. The area shown in the inset of Figure 2.3 is the problem domain. Note that $100 \leq x \leq 400$ m and $0 \leq y \leq 300$ m. This area is located near a pumping well as shown in Figure 2.3. We consider a numerical model of the entire problem domain in Chapter 3, when we introduce a sink term to simulate discharge from the pumping well. For now, suppose that observation wells are located around the section of the problem domain shown in the inset. Water levels are measured in each observation well and are given in Figure 2.3. We have now specified the boundary

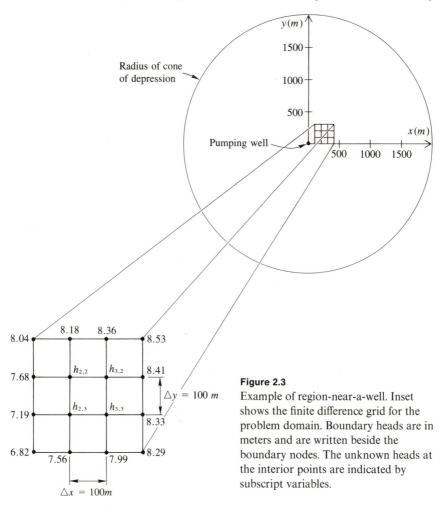

Figure 2.3
Example of region-near-a-well. Inset shows the finite difference grid for the problem domain. Boundary heads are in meters and are written beside the boundary nodes. The unknown heads at the interior points are indicated by subscript variables.

conditions for the problem, and we can use Equation 2.5 to solve for the heads at the four interior points of Figure 2.3, namely, (2, 2), (2, 3), (3, 2), and (3, 3).

Finite difference equations need to be written for the four unknowns, $h_{2,2} = h(200, 200)$, $h_{2,3} = h(200, 100)$, $h_{3,2} = h(300, 200)$, and $h_{3,3} = h(300, 100)$. Recall the convention that the first subscript refers to column number and the second subscript refers to row number. The numbers in parentheses refer to the x and y coordinates. By applying Equation 2.5 to each interior node, we obtain the following set of algebraic equations:

$$\text{At } i = 2, j = 2: \quad \underline{h_{1,2}} + h_{3,2} + \underline{h_{2,1}} + h_{2,3} - 4h_{2,2} = 0$$

$$\text{At } i = 2, j = 3: \quad \underline{h_{1,3}} + h_{3,3} + h_{2,2} + \underline{h_{2,4}} - 4h_{2,3} = 0$$

$$\text{At } i = 3, j = 2: \quad h_{2,2} + \underline{h_{4,2}} + \underline{h_{3,1}} + h_{3,3} - 4h_{3,2} = 0$$

$$\text{At } i = 3, j = 3: \quad h_{2,3} + \underline{h_{4,3}} + h_{3,2} + \underline{h_{3,4}} - 4h_{3,3} = 0$$

where boundary heads are underlined.

Note that the points at the corners of the problem domain are not used. Because the heads along the boundaries are given, the four equations contain only four unknowns. In Problem 2.1, readers are asked to find the solution to these four equations by simultaneous solution. A particular analytical solution of this regional problem, which includes the pumping well, is

$$h(x, y) - 10 = 1.06 \ln\left(\frac{\sqrt{x^2 + y^2}}{2000}\right) \tag{2.6}$$

and is discussed in more detail in Chapter 3, when the regional problem is solved. In Problem 2.1, readers are asked to use this analytical solution to check the validity of the numerical solution for the heads in the Dirichlet problem considered here. In the next section, we describe the use of iterative methods for solving a set of linear equations such as the ones just given.

2.3   ITERATIVE METHODS

If the finite difference equation, Equation 2.5, were solved for $h_{i,j}$, then

$$h_{i,j} = \frac{h_{i-1,j} + h_{i+1,j} + h_{i,j-1} + h_{i,j+1}}{4} \tag{2.7}$$

That is, the value of $h_{i,j}$ at any point is the average value of head computed from its four nearest neighbors in the nodal array. Equation 2.7 is often called

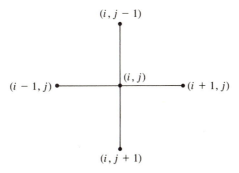

$(i, j - 1)$

$(i - 1, j)$ (i, j) $(i + 1, j)$

$(i, j + 1)$

Figure 2.4
Five-point star representing finite difference approximation
for Laplace's equation.

the five-point operator because the algebraic equations which approximate Laplace's equation are created one after another by moving the star of five points (Figure 2.4) throughout the domain of the problem.

In the previous section, the Dirichlet boundary value problem involving Laplace's equation could be solved directly as a set of four simultaneous linear equations. However, in problems having a large number of nodes, simultaneous solutions by hand are impractical. Instead, suppose we guess a set of trial answers and then successively improve our guesses until we get the right answer. In short, iterative methods consist of guessing and adjusting. We shall discuss three iterative techniques—Jacobi iteration, Gauss–Seidel iteration, and successive over relaxation (SOR). Jacobi iteration is the least efficient and is seldom used. Gauss–Seidel iteration can be considered to be a special case of successive over relaxation. Successive over relaxation is generally the most efficient for groundwater problems.

Jacobi Iteration

We illustrate iterative techniques by applying them to the region-near-a-well problem of Section 2.2. We use Equation 2.7 and apply it to each interior node in Figure 2.3. We are not concerned about the order in which we do the iterations, because we are not using newly computed values of head to compute the head at an adjacent node. If we let m be the iteration index, then the formula we will use is

$$h_{i,j}^{m+1} = \frac{h_{i-1,j}^m + h_{i+1,j}^m + h_{i,j-1}^m + h_{i,j+1}^m}{4} \qquad (2.8)$$

First, we must guess a set of trial answers ($m = 1$) to begin the iteration process. For $m = 1$, we use the values $h_{2,2}^1 = 8$, $h_{3,2}^1 = 8.5$, $h_{2,3}^1 = 7$, and $h_{3,3}^1 = 8$. According to Equation 2.8, the values of head for $m = 2$ are $h_{2,2}^2 = 7.84$, $h_{3,2}^2 = 8.19$, $h_{2,3}^2 = 7.69$, and $h_{3,3}^2 = 7.96$. These values are used to solve for the head values at the next iteration level ($m = 3$). Iteration continues until the solution converges, that is, until the difference between the answers generated at two successive iteration levels is less than a preset value known as the error tolerance or convergence criterion.

Gauss–Seidel Iteration

In this method, we work through the grid in an orderly way; that is, we start at $i = 2$ and $j = 2$ and sweep across from left to right and down line-by-line as if we were reading a page (Figure 2.5). In this way, we can always use two newly computed values in the iteration formula. Thus the Gauss–Seidel iteration formula is

$$h_{i,j}^{m+1} = \frac{h_{i-1,j}^{m+1} + h_{i,j-1}^{m+1} + h_{i+1,j}^m + h_{i,j+1}^m}{4} \tag{2.9}$$

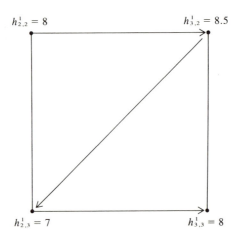

$h_{2,2}^1 = 8$ $h_{3,2}^1 = 8.5$

$h_{2,3}^1 = 7$ $h_{3,3}^1 = 8$

Figure 2.5
Gauss–Seidel iteration. The arrows indicate the order of iteration. Initial guesses for unknown heads h_{ij}^m are written beside each interior node for the region-near-a-well example. The superscript indicates the iteration level.

For example, to apply Gauss–Seidel iteration to the region-near-a-well example, we would first compute $h_{2,2}^2$ using values of h at the $m = 1$ iteration level for the nodes to the right and below node $(2, 2)$, and known boundary heads for the other two nodes in the iteration formula, Equation 2.9. Next, $h_{3,2}^2$ is computed using $h_{2,2}^2$ and $h_{3,3}^1$ as well as two boundary heads. Note that we are able to use the newly computed value of head at $(2, 2)$ when computing $h_{3,2}^2$. In general, for problems with a larger number of interior nodes, heads at the nodes to the left of node (i, j) and above node (i, j) will be computed before the head at node (i, j). In other words, $h_{i-1,j}$ and $h_{i,j-1}$ will be known at the $m + 1$ iteration level when it is time to evaluate $h_{i,j}$ at the $m + 1$ iteration level. The use of newly computed values whenever possible makes Gauss–Seidel iteration more efficient than Jacobi iteration.

Successive Over Relaxation (SOR)

The change between two successive Gauss–Seidel iterations is called the residual c. The residual c is then defined by

$$c = h_{i,j}^{m+1} - h_{i,j}^m \qquad (2.10)$$

By replacing $h_{i,j}^m$ with $h_{i,j}^{m+1}$ after each calculation, the Gauss–Seidel procedure liquidates or relaxes the residuals at every node and thus leads to a solution of each algebraic equation.

In the method of SOR, the Gauss–Seidel residual is multiplied by a relaxation factor ω where $\omega \geq 1$, and the new value $h_{i,j}^{m+1}$ is given by the formula

$$h_{i,j}^{m+1} = h_{i,j}^m + \omega c \qquad (2.11)$$

That is, the SOR equation for updating $h_{i,j}$ at the $(m + 1)$ iteration is obtained by substituting the right-hand side of Equation 2.9 into Equation 2.10 and the result into Equation 2.11.

$$h_{i,j}^{m+1} = (1 - \omega)h_{i,j}^m + \omega \frac{h_{i-1,j}^{m+1} + h_{i,j-1}^{m+1} + h_{i+1,j}^m + h_{i,j+1}^m}{4} \qquad (2.12)$$

In SOR, more residual is added to $h_{i,j}^m$ than in the Gauss–Seidel method, and $h_{i,j}^{m+1}$, as computed by SOR, is said to be over relaxed. If $0 < \omega < 1$, the updated head value is said to be under relaxed. If $\omega = 1$, the SOR iteration formula, Equation 2.12, reduces to the Gauss–Seidel formula, Equation 2.9.

There are methods of selecting the best value of ω for a particular problem (see, for example, Remson et al., 1971), but, because of the complexity of the procedure, it is easier to find the optimum ω by trial and error. In general, $1 \leq \omega \leq 2$.

The relaxation formula, Equation 2.12, can be viewed as an interpolation formula in the case of under relaxation ($0 < \omega < 1$) and as an extrapolation formula in the case of over relaxation ($\omega > 1$). First, we can see this for under relaxation if we write Equation 2.11 as

$$h_{i,j}^{m+1} = (1 - \omega)h_{i,j}^m + \omega \tilde{h}_{i,j}^{m+1} \tag{2.13}$$

where

$$\tilde{h}_{i,j}^{m+1} = \frac{h_{i-1,j}^{m+1} + h_{i,j-1}^{m+1} + h_{i+1,j}^m + h_{i,j+1}^m}{4} \tag{2.14}$$

For $0 < \omega < 1$, Equation 2.13 represents a weighted average of $h_{i,j}^m$ and $\tilde{h}_{i,j}^{m+1}$. Therefore, the computed next guess $h_{i,j}^{m+1}$ using under relaxation must lie between the old guess $h_{i,j}^m$ and the Gauss–Seidel value $\tilde{h}_{i,j}^{m+1}$. Although the Gauss–Seidel value $\tilde{h}_{i,j}^{m+1}$ will reduce the residual to zero for one equation of the system, the newly calculated $\tilde{h}_{i,j}^{m+1}$ will ripple throughout the system of equations by affecting other residuals. By choosing the new guess $h_{i,j}^{m+1}$ to go only some fraction of the distance between $h_{i,j}^m$ and the Gauss–Seidel new guess $\tilde{h}_{i,j}^{m+1}$, we are being more cautious. We tamper less with the other equations which contain $h_{i,j}^{m+1}$.

For over relaxation ($\omega > 1$), on the other hand, we are so bold as to extrapolate past the Gauss–Seidel value $\tilde{h}_{i,j}^{m+1}$. We can rearrange Equation 2.13 to show how the new guess $h_{i,j}^{m+1}$ is a function of the relaxation parameter ω.

$$h_{i,j}^{m+1} = h_{i,j}^m + (\tilde{h}_{i,j}^{m+1} - h_{i,j}^m)\omega \tag{2.15}$$

Equation 2.15 represents a straight line (Figure 2.6) when $h_{i,j}^{m+1}$ is plotted in terms of ω. From the figure, we can see that, if the previous guess $h_{i,j}^m$ were still far from the answer, then a large value of ω would bootstrap us there more quickly. But, as we get close to the answer, a big value of ω would tend to make us overshoot the answer. That is the reason that values of ω between 1 and 2 tend to be optimum. They lead to steps bold enough to get us quickly toward the answer, but not so bold as to walk past the solution.

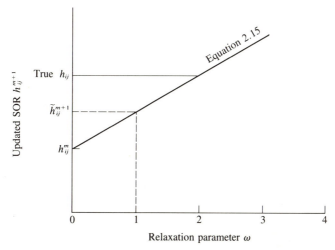

Figure 2.6
Graphic interpretation of under relaxation ($0 < \omega < 1$) as interpolation between old value of head h_{ij}^m and the Gauss–Seidel value $\tilde{h}_{ij}^{m+1}$ and of over relaxation ($\omega > 1$) as extrapolation of the Gauss–Seidel value.

2.4 GAUSS–SEIDEL COMPUTER PROGRAM

The iterative procedures for solving algebraic equations are well suited to computer solution. We present a short FORTRAN program (Figure 2.7) to solve the region-near-a-well problem (Section 2.2) by Gauss–Seidel iteration. The program is readily generalized to SOR and to problems containing many more nodes than in this simple example. We use the index convention where I represents the column number from left to right and J represents the row number from top to bottom. The I and J of the FORTRAN program refer to the i and j of Figure 2.2.

The program can be divided into three parts—initialization (lines 1 to 9), computation (lines 10 to 23), and printout (lines 24 to 28).

The boundary values, as well as the initial guesses for the four interior values of head, are set in lines 1 to 9. Line 13 is a counter of the number of iterations. The most important portion of the program is contained in lines 14 to 23. The *DO* loops move the five-point operator systematically through the interior points. Line 18 is the FORTRAN expression for Equation 2.9. An iteration index is not needed in the FORTRAN expression because the computer automatically replaces old values in the head array with new values as soon as they are computed. In line 23, the program tests for convergence. For this example,

Figure 2.7
Computer program using Gauss–Seidel iteration for the region-near-a-well example.

```
 1.   C  GAUSS-SEIDEL ITERATION FOR REGION NEAR A WELL
 2.         DIMENSION H(4,4)
 3.   C  BOUNDARY VALUES -- UPPER, LOWER, LEFT, RIGHT
 4.         DATA H(1,1),H(2,1),H(3,1),H(4,1)/8.04,8.18,8.36,8.53/
 5.         DATA H(1,4),H(2,4),H(3,4),H(4,4)/6.82,7.56,7.99,8.29/
 6.         DATA H(1,2),H(1,3)/7.68,7.19/
 7.         DATA H(4,2),H(4,3)/8.41,8.33/
 8.   C  INITIAL GUESSES FOR INTERIOR POINTS
 9.         DATA H(2,2),H(3,2),H(2,3),H(3,3)/8.,8.5,7.,8./
10.   C  KEEP TRACK OF NUMBER OF ITERATIONS AND OF LARGEST ERROR
11.         NUMIT = 0
12.   35    AMAX = 0.
13.         NUMIT = NUMIT + 1
14.   C  SWEEP INTERIOR POINTS WITH 5-POINT OPERATOR
15.         DO 40 J=2,3
16.         DO 40 I=2,3
17.         OLDVAL = H(I,J)
18.         H(I,J) = (H(I-1,J) + H(I+1,J) + H(I,J-1) + H(I,J+1))/4.
19.         ERR = ABS(H(I,J) - OLDVAL)
20.         IF(ERR.GT.AMAX) AMAX = ERR
21.   40    CONTINUE
22.   C  DO ANOTHER ITERATION IF LARGEST ERROR AFFECTS 2ND DECIMAL PLACE
23.         IF(AMAX.GT.0.01) GO TO 35
24.   C  WE ARE DONE
25.         PRINT 50,NUMIT,((H(I,J),I=1,4),J=1,4)
26.   50    FORMAT(///1X,'NUMBER OF ITERATIONS IS ',I4,///4 (4F10.2//))
27.         STOP
28.         END
```

NUMBER OF ITERATIONS IS   4

8.04	8.18	8.36	8.53
7.68	7.93	8.19	8.41
7.19	7.68	8.05	8.33
6.82	7.56	7.99	8.29

Figure 2.8
Solution for the region-near-a-well example. Heads are printed in the same configuration as the nodes in Figure 2.3.

the solution converges when all the unknown head values change by less than 0.01 m from one iteration to the next. When the convergence criterion is satisfied, the program prints out both boundary and interior heads (line 25). The output is shown in Figure 2.8.

2.5 BOUNDARY CONDITIONS

The Gauss–Seidel program in Section 2.4 was for a Dirichlet problem in which all the boundary heads were known. In this section, we illustrate how to solve a mixed boundary condition problem numerically by considering the model of a regional groundwater flow system described in Section 1.4. The mathematical description of the problem was given in Figure 1.7. Three of the boundaries are described by Neumann-type boundary conditions. Physically they are no-flow boundaries. No flow crosses the vertical lines of symmetry which mark the groundwater divides, and no flow crosses the bottom boundary which represents the contact between permeable and impermeable rock. How do we handle the conditions $\partial h/\partial x = 0$ and $\partial h/\partial y = 0$ along the no-flow boundaries? Since the unknowns are the heads $h_{i,j}$, we would like to change the conditions on the first derivatives of head into conditions on the heads themselves.

Let us consider the left boundary. The finite difference approximation to $\partial h/\partial x$ is

$$\frac{\partial h}{\partial x} \simeq \frac{h_{i+1,j} - h_{i-1,j}}{2\Delta x} \tag{2.16}$$

For the left boundary, the point referred to by indices $(i + 1, j)$ is inside the problem domain, but the point referred to by indices $(i - 1, j)$ is outside. Therefore, we expand the finite difference problem domain by one additional column to the left by putting in a column of so-called imaginary or fictitious nodes. The boundary condition $\partial h / \partial x = 0$ translates in finite difference form to $h_{i+1, j} = h_{i-1, j}$. That is, the value of head along the fictitious column must reflect across the left boundary. Thus, the left boundary is a symmetry line. The right and bottom no-flow boundaries are also handled by creating additional fictitious nodes. The top boundary has specified heads and, therefore, no fictitious nodes are needed along this boundary.

Let us consider a specific numerical example. In Figure 1.7, let $s = 200$ m, $y_0 = 100$ m, and $c = 0.02$. We divide the rectangular problem domain $0 \leq x \leq s$ and $0 \leq y \leq y_0$ into an eleven by six grid of points, where eleven is the number of columns in the x direction and six is the number of rows in the y direction. We use the same indexing convention whereby I represents column number from left to right and J represents row number from top to bottom. The addition of fictitious nodes for the no-flow boundaries leads to a total of thirteen columns and seven rows (Figure 2.9). The spacing between nodes is $\Delta x = \Delta y = 20$ m. The FORTRAN program and output are shown in Figures 2.10 and 2.11, respectively.

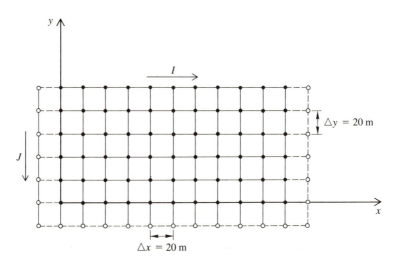

Figure 2.9
Finite difference grid for computer model of the regional groundwater flow example. Solid circles represent the problem domain of Figure 1.7. Open circles represent fictitious nodes used to specify no-flow boundary conditions. The fictitious node at the upper left-hand corner of the diagram represents $I = 1, J = 1$.

The boundary values and initial guesses for the unspecified heads are set in lines 1 to 26. The Gauss–Seidel iteration is done in lines 27 to 34. Readers should compare this program with the one for the region-near-a-well example in Figure 2.7. The structure of both programs is quite similar. The major difference is that, in the current example, the no-flow boundary conditions (lines 18 to 26) need to be enforced after each iteration. Lines 27 to 36 in Figure 2.10 perform exactly the same functions as lines 14 to 23 in Figure 2.7. However, the convergence criterion in Figure 2.10 is that the change in head at every node be less than 0.001 m between successive iterations. Readers can test that a less stringent convergence criterion can lead to unacceptable answers.

The output in Figure 2.11 can be contoured, and some regional flow lines can be drawn perpendicular to the equipotential lines. Thus the nature of a regional flow system has been explored with a very simple computer program. The numerical answers in this case could be compared with the analytical solution, Equation 2.1.

Figure 2.10
Computer program for regional flow example using Gauss–Seidel iteration.

```
1.    C   REGIONAL FLOW SYSTEM EXAMPLE
2.            DIMENSION H(13,7)
3.    C   INITIALIZE ALL H(I,J) VALUES TO BE 100.
4.            DO 5 J=1,7
5.            DO 5 I=1,13
6.            H(I,J) = 100.
7.        5   CONTINUE
8.    C   WATER TABLE BOUNDARY
9.            DX=20.
10.           DO 10 I=2,12
11.           H(I,1) = 0.02*DX*(I-2)+100.
12.      10   CONTINUE
13.   C   KEEP TRACK OF NUMBER OF ITERATIONS AND OF LARGEST ERROR
14.   C   NO-FLOW BOUNDARIES NEED TO BE RESET WITHIN EACH ITERATION LOOP
15.           NUMIT = 0
16.      35   AMAX = 0.
17.           NUMIT = NUMIT + 1
18.   C   LEFT AND RIGHT NO-FLOW BOUNDARIES
19.           DO 20 J=1,7
20.           H(1,J) = H(3,J)
21.           H(13,J) = H(11,J)
22.      20   CONTINUE
```

```
      C   BOTTOM NO-FLOW BOUNDARY
            DO 30 I=2,12
            H(I,7) = H(I,5)
   30       CONTINUE
      C   SWEEP INTERIOR POINTS WITH 5-POINT OPERATOR
            DO 40 J=2,6
            DO 40 I=2,12
            OLDVAL = H(I,J)
            H(I,J) = (H(I-1,J) + H(I+1,J) + H(I,J-1) + H(I,J+1))/4.
            ERR = ABS(H(I,J) - OLDVAL)
            IF(ERR.GT.AMAX) AMAX=ERR
   40       CONTINUE
      C   DO ANOTHER ITERATION IF LARGEST ERROR AFFECTS 3RD DECIMAL PLACE
            IF(AMAX.GT.0.001) GO TO 35
      C   WE ARE DONE.
            PRINT 50,NUMIT,((H(I,J),I=2,12),J=1,6)
   50       FORMAT(///1X,'NUMBER OF ITERATIONS IS',I4,///6(11F8.2///))
            STOP
            END
```

[35]

Figure 2.11
Output for regional flow example from the computer program in Figure 2.10. Output consists of values of head for nodes represented by solid circles in Figure 2.9; head values for fictitious nodes are not printed. Equipotential lines and flow paths can be drawn over the output.

NUMBER OF ITERATIONS IS 109

100.00	100.40	100.80	101.20	101.60	102.00	102.40	102.80	103.20	103.60	104.00
100.63	100.78	101.03	101.34	101.66	101.99	102.33	102.65	102.95	103.21	103.35
100.98	101.05	101.22	101.45	101.71	101.99	102.26	102.53	102.76	102.93	103.00
101.17	101.22	101.35	101.53	101.75	101.98	102.22	102.44	102.62	102.74	102.79
101.28	101.32	101.42	101.58	101.77	101.98	102.19	102.38	102.54	102.64	102.68
101.31	101.35	101.45	101.60	101.78	101.98	102.18	102.36	102.51	102.61	102.64

Notes and Additional Reading

1. Bennett (1976, pp. 119–135 of Part IV, Finite Difference Methods) and Remson et al. (1971, pp. 63–67) give additional details on the concept of discretization and the definition of finite differences. Remson et al. also provide additional information on iteration and convergence (pp. 177–188) and the method of successive over relaxation (pp. 189–203). Neumann-type boundary conditions are covered in more detail on pp. 90–92 of Remson et al. (1971).

2. Several researchers have used successive over relaxation to solve groundwater flow problems governed by a general form of the steady-state flow equation in which the aquifer is allowed to be heterogeneous and anisotropic:

$$\frac{\partial}{\partial x}\left(K_x \frac{\partial h}{\partial x}\right) + \frac{\partial}{\partial y}\left(K_y \frac{\partial h}{\partial y}\right) = 0 \tag{2.17}$$

This equation is similar to Equation 1.12, but in Equation 2.17 the aquifer is anisotropic as well as heterogeneous. Anisotropy requires that K be a tensor as discussed in Appendix A. In the notation of Appendix A, $K_x = K_{11}$ and $K_y = K_{22}$ when $K_{12} = K_{21} = 0$.

   An application of this governing equation, Equation 2.17, to a groundwater lake system is presented by McBride and Pfannkuch (1975), who use successive over relaxation to solve the numerical model. Freeze and Witherspoon (1966, 1967, 1968) use the method of successive over relaxation to solve a model of a regional flow system allowing for anisotropic and heterogeneous conditions. They also use Equation 2.17 as the governing equation. Their work is a logical extension of Toth (1962, 1963).

3. In Section 2.5, we introduced a technique for handling specified flow boundary conditions through the use of fictitious nodes. Another convention can be used for block-centered nodes. In this case, the boundary is placed between the fictitious node and the neighboring node within the problem domain. The finite difference expression for flow across a right-hand side boundary in the x direction using this convention is

$$\frac{h_{i+1,j} - h_{i,j}}{\Delta x} = \frac{-q_x}{K} \tag{2.18}$$

For no-flow conditions, Equation 2.18 simplifies to

$$h_{i+1,j} = h_{i,j}$$

Problems

2.1 Substitute the known boundary head values given in Figure 2.3 into the four finite difference equations in Section 2.2. Then solve the four equations simultaneously for the unknown values of head, that is, $h_{2,2}$, $h_{2,3}$, $h_{3,2}$, and $h_{3,3}$.

2.2 (a) Solve the region-near-a-well problem in Section 2.2 using Gauss–Seidel iteration. Using hand calculations, sweep through the four unknown, interior points several times until the answers no longer change in the second decimal place. The initial guesses, the heads at $m = 1$, are given in Figure 2.5. You will see that, as the iteration procedure progresses, the values change less between successive iterations. This fact means that the solution is converging. Note that the head values after the first Gauss–Seidel iteration are closer to the correct answers than those computed after one Jacobi iteration (see Section 2.3).

(b) Compare the Gauss–Seidel answers with the answers you generated by solving the four simultaneous equations directly in Problem 2.1.

(c) Compare your answers with the analytical solution, Equation 2.6.

(d) What is the error in the finite difference answers? How could the error be reduced?

(e) The initial guesses used in part (a) were good ones. Only six iterations were needed to obtain answers accurate to two decimal places. Suppose you initially guessed a head value of 10 m for each interior point. Do several Gauss–Seidel iterations to convince yourself that the solution converges to the correct answer despite the choice of the initial guess.

2.3 Modify the program in Figure 2.7 to solve Laplace's equation using successive over relaxation (SOR) and compare the results with those in Figure 2.8. You will see that, for this particular problem, more iterations are required for SOR than for Gauss–Seidel iteration. Refer to Figure 2.6 and to Problem 2.2e and offer an explanation.

2.4 Change the upper boundary condition in the program in Figure 2.10 to $h(x, y_0) = y_0 + x \tan \alpha + a\{[\sin(bx/\cos \alpha)]/\cos \alpha\}$ where $\alpha = 1.1°$ (or 0.0192 rad) and $b = 2\pi/\lambda$ where $\lambda = 80$ m and $a = 2.0$ m. This boundary condition is the one considered by Toth (1963). The parameters a, α, and λ are defined in Figure 2.12. Sketch approxi-

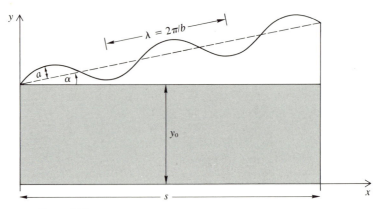

Figure 2.12
Regional groundwater flow system with sinusoidal water table.

mate equipotential lines and flow paths on the computer output, and compare them with the results for a linear water table (Figure 2.11).

2.5 Write a finite difference expression for d^2h/dx^2 at point i for the following arrangement of nodal points in which the nodal spacing is variable.

CHAPTER 3

Finite Differences: Steady-State Flow (Poisson's Equation)

3.1 INTRODUCTION

In Chapter 2, we described a finite difference model based on Laplace's equation. In this chapter, we extend the treatment of the finite difference technique to allow consideration of sources and sinks through the use of Poisson's equation. We also consider treatment of unconfined flow under the Dupuit assumptions. We illustrate each of these topics by comparing a classical analytical solution with the finite difference solution of a similar problem. Specifically, the examples consider areal recharge from precipitation, drawdowns around a discharging well, and seepage through a dam.

3.2 POISSON'S EQUATION

Recharge from precipitation and discharge from a well are examples of a distributed source and point sink of groundwater, respectively. The addition or withdrawal of groundwater means that the divergence, or volume rate of outflow

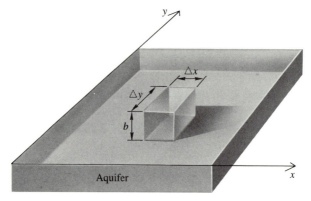

Figure 3.1
Two-dimensional horizontal aquifer containing a representative
elemental volume for derivation of Poisson's equation.

per unit aquifer volume, is not identically zero everywhere in the problem
domain, and hence Laplace's equation is no longer the governing equation.

Let us consider a two-dimensional, horizontal aquifer whose thickness is
uniformly of height b (Figure 3.1). The first step in the derivation of a new
governing equation is to do a continuity analysis similar to that in Section 1.2.
We consider the elemental volume within the aquifer shown in Figure 3.1. Let
$R(x, y)$ be the volume of water added per unit time per unit aquifer area to the
infinitesimal volume around the point (x, y). For steady-state conditions, the
volume rate of outflow must equal $R(x, y)\,\Delta x\,\Delta y$. That is,

$$\frac{\partial q_x}{\partial x}\,\Delta x(b\,\Delta y) + \frac{\partial q_y}{\partial y}\,\Delta y(b\,\Delta x) = R(x, y)\,\Delta x\,\Delta y \tag{3.1}$$

We use Darcy's law to substitute for q_x and q_y. Also, we define the transmissivity
T to be the product of hydraulic conductivity K and aquifer thickness b. Making
these substitutions and dividing through by $-T\,\Delta x\,\Delta y$ yields Poisson's equation

$$\frac{\partial^2 h}{\partial x^2} + \frac{\partial^2 h}{\partial y^2} = -\frac{R(x, y)}{T} \tag{3.2}$$

If $R(x, y)$ is equal to zero everywhere in the problem domain, then Poisson's
equation reduces to Laplace's equation. The term $R(x, y)$ is used to simulate
both distributed and point sources (positive values) and sinks (negative values)
and has units of length per unit time.

3.3 ISLAND RECHARGE

Jacob (1943) considers the one-dimensional version of Equation 3.2, with R equal to a constant, to simulate groundwater flow beneath Long Island, New York (Figure 3.2). The governing equation is

$$\frac{d^2h}{dx^2} = -\frac{R}{T} \tag{3.3}$$

Jacob uses this governing equation to simulate an unconfined aquifer by assuming that the actual saturated thickness $h + b$ is approximated by the distance b. That is, he assumes that $b \gg h$. We consider unconfined flow later in the chapter, but for now let us accept Jacob's assumption. It is also assumed in the model that the water table intersects sea level at the edges of the island and that the water table profile is symmetric across the island. Therefore, we need to consider only one-half of the problem domain—either $0 \le x \le \ell$ or $-\ell \le x \le 0$. If we consider the right-hand side of the problem domain, the boundary conditions are $h = 0$ at $x = \ell$ and $dh/dx = 0$ at $x = 0$. The model can be solved analytically by integrating twice with respect to x. The general solution is

$$h(x) = -\frac{R}{T}\frac{x^2}{2} + a_1 x + a_2 \tag{3.4}$$

where a_1 and a_2 are arbitrary constants which can be determined from the boundary conditions. The boundary condition at $x = 0$ is used to find that

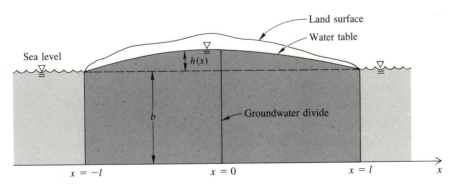

Figure 3.2
One-dimensional model (Jacob, 1943) of Long Island, New York.

$a_1 = 0$, and the boundary condition at $x = \ell$ is used to find that $a_2 = R\ell^2/2T$. Therefore, the solution to the model is

$$h(x) = \frac{R}{2T}(\ell^2 - x^2) \tag{3.5}$$

3.4   FINITE DIFFERENCE MODELS

Island Recharge

We return to the two-dimensional form of Poisson's equation by considering an island twice as long as it is wide (Figure 3.3). Suppose we know the transmissivity T of the aquifer and the elevation of the water level in the well at the center of the island. The recharge rate R to the water table is not directly measurable in the field. However, the recharge rate can be estimated by using a mathematical model based upon Poisson's equation and by using the one data point of head at the center of the island. The use of Poisson's equation carries the assumption that the thickness of the aquifer is much larger than the head relative to sea level.

The boundary conditions can be treated in two ways. The simplest way, but wasteful of computer memory, is to consider the entire island to be the problem domain with all boundary value heads equal to zero. The second way is to use no-flow boundary conditions along symmetry lines (the x, y axes in Figure 3.3)

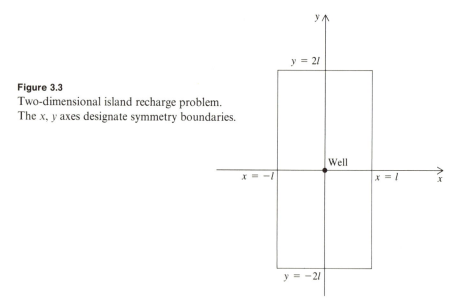

Figure 3.3
Two-dimensional island recharge problem.
The x, y axes designate symmetry boundaries.

so that only one-quarter of the island need make up the problem domain. Depending on which method of handling boundary conditions is chosen, we can modify either of our computer programs, Figure 2.7 or Figure 2.10, to solve the recharge problem. The major change we need to make is that line 18 in Figure 2.7 or line 31 in Figure 2.10 must be modified to solve Poisson's equation instead of Laplace's equation. If second differences are substituted for the second partial derivatives, then the finite difference form of Poisson's equation is

$$\frac{h_{i-1,j} - 2h_{i,j} + h_{i+1,j}}{(\Delta x)^2} + \frac{h_{i,j-1} - 2h_{i,j} + h_{i,j+1}}{(\Delta y)^2} = -\frac{R}{T} \qquad (3.6)$$

The FORTRAN statement which automatically incorporates Gauss–Seidel iteration for Equation 3.6 is

$$H(I, J) = (H(I - 1, J) + H(I + 1, J) + H(I, J - 1)$$
$$+ H(I, J + 1) + DX*DX*R/T)/4 \qquad (3.7)$$

where DX is the nodal spacing $\Delta x = \Delta y$.

Equation 3.7, however, requires that a recharge rate R be specified; yet R is what we are trying to determine. Therefore, we must find R by trial and error. First, we guess a value of R and use the computer program to generate the head array including the head at the center of the island. Next, we compare the head at the center of the island computed by the model with the measured head. If the computed head is much different from the measured value, we can adjust the value of R and let the program generate another set of heads until the computed head at the center of the island is close to the measured value. The trial-and-error procedure requires an initial guess for R. The analytical solution, Equation 3.5, for the one-dimensional problem should provide a good first approximation for R. In Problem 3.2, we consider the solution of this problem using specific values for T, ℓ, and the water level in the well at the center of the island.

If we compute $h(x, y)$ given values of T and R, we are working in what is called the forward direction. In Problem 3.2, you are asked to compute R given the head at one point. This problem is the inverse of the forward problem. The advantage of the analytical solution for the one-dimensional problem is that the inverse problem can be solved by algebra, and no trial-and-error procedure is required. But, in the two-dimensional case, the inverse problem is solved by solving the forward problem many times with improved guesses for R.

Another kind of inverse problem involves finding values for transmissivity or hydraulic conductivity given measured values of the heads. This inverse problem could also be solved by trial and error. However, it has been demonstrated that several different transmissivity distributions may generate the same head distribution because heads are not particularly sensitive to changes in transmissivity (Gillham and Farvolden, 1974). Several researchers are currently attempting to formulate more efficient ways of solving the inverse problem for transmissivities. For details see Cooley (1977, 1979), Neuman (1973), Neuman and Yakowitz (1979), Neuman et al. (1980), and Neuman (1980).

Well Drawdown (Confined Aquifer)

Sources and sinks are mathematically represented in Poisson's equation by the recharge function $R(x, y)$. We want to use Poisson's equation to describe discharge from a well that fully penetrates a horizontal, isotropic aquifer of thickness b. If we assume the well is screened throughout the entire thickness of the aquifer, then flow will be horizontal and two-dimensional. Because the well is located at a single point (x_0, y_0), it is referred to as a point sink (or source, if it is an injection well), whereas areal recharge from precipitation is a distributed source.

In the mesh-centered grid shown in Figure 3.4, the axis of the well is presumed to be centered within an infinitesimal volume $b \Delta x \Delta y$ such as that shown in Figure 3.1. Suppose the well is pumped at a rate Q, where Q has units of volume/time. Then, the recharge function for the infinitesimal surface area about (x_0, y_0) is

$$R(x_0, y_0) = \frac{-Q}{\Delta x \, \Delta y} \tag{3.8}$$

because $R(x, y)$ is the volume of recharge per unit time per unit aquifer area. The negative sign exists in Equation 3.8 because of the convention we established when deriving Poisson's equation (Equation 3.2) that withdrawals from the aquifer (sinks) are negative. $R(x, y)$ is equal to zero outside the infinitesimal volume containing the well. Thus we have completely defined $R(x, y)$ throughout the problem domain and have mathematically described discharge from a fully penetrating well in terms of Poisson's equation.

This mathematical description of $R(x, y)$ for a well is readily incorporated into a finite difference computer program. Suppose the well pumps 2000 m^3 day^{-1} in an aquifer whose transmissivity is 300 m^2 day^{-1}. Before pumping, the static water level in the well is measured to be 10 m. We make the assumption that drawdown extends out to a radial distance of 2000 m from the well (Figure 2.3). That is, the static water level remains unaffected for distances greater than 2000 m from the well.

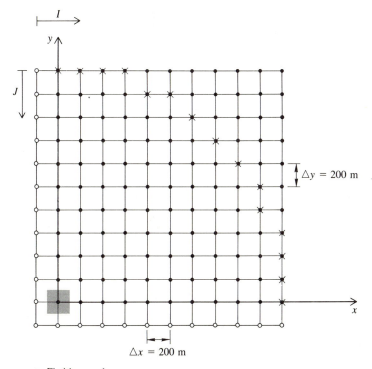

$\Delta y = 200 \text{ m}$

$\Delta x = 200 \text{ m}$

○ Fictitious node

✕ Specified head node

Figure 3.4
Finite difference grid for the drawdown example. The problem domain is the upper
right-hand portion of the cone of depression shown in Figure 2.3. Shaded cell
represents the well at the origin. Open circles are fictitious nodes that are used to
simulate no-flow boundaries along the x and y axes. Head is equal to 10 m at the edge
of the cone of depression. These specified head nodes are shown by the symbol ✕.

After pumping starts, equipotential lines will be circles about the well and
flow lines will be radial. Although the problem has cylindrical symmetry, we
use rectangular coordinates for the finite difference formulation.

Because of symmetry, we may consider just the one quadrant shown in
Figure 3.4 and require that the x and y symmetry axes be no-flow boundaries.
The outer boundary at a radius of 2000 m is circular, and the head along it is
10 m. The spacing between nodes is chosen to be $\Delta x = \Delta y = 200$ m. The
finite difference grid is shown in Figure 3.4. In the computer program (Figure
3.5), the circular boundary is approximated in lines 17 to 30 by finding that

Figure 3.5
Finite difference program for solving the drawdown example.

```
1.       C     DRAWDOWN EXAMPLE - CONFINED AQUIFER
2.             DIMENSION H(12,12)
3.             DIMENSION KBEG(12)
4.             OMEGA=1.8
5.             N=10
6.             N1=N+1
7.             N2=N+2
8.       C     INPUT VARIABLES
9.             DX=200.
10.            RWELL=-2000./DX/DX
11.            T=300.
12.      C     SET INITIAL GUESSES FOR H(I,J)
13.            DO 5 J=1,N2
14.            DO 5 I=1,N2
15.            H(I,J)=5.
16.      5     CONTINUE
17.      C     SET STATIC WATER LEVEL BOUNDARY CONDITION AT NODE IN EACH COLUMN
18.      C     WHICH IS CLOSEST TO BEING 2000-M FROM THE ORIGIN
19.            DO 25 I=2,N2
20.            DMIN=20000.
21.            DO 20 J=1,N1
22.            X=DX*(I-2)
23.            Y=2000.-DX*(J-1)
24.            DIST=SQRT(X*X+Y*Y)
25.            DELT=ABS(DIST-2000.)
26.            IF(DELT.GE.DMIN)GO TO 20
27.            DMIN=DELT
28.            KBEG(I)=J+1
29.      20    CONTINUE
30.      25    CONTINUE
```

```
31.    C    SET THE STATIC WATER LEVEL AT ALL NODES FARTHER
32.    C    THAN 2000-M OUT TO THE CORNERS, EQUAL TO 10. METERS
33.              DO 32 I=2,N2
34.              JBND=KBEG(I)-1
35.              DO 32 J=1,JBND
36.              H(I,J)=10.
37.    32        CONTINUE
38.              NUMIT=0
39.    35        AMAX=0.
40.              NUMIT=NUMIT+1
41.              IF(NUMIT.GT.2000)GO TO 45
42.    C    LEFT AND BOTTOM SYMMETRY NO-FLOW BOUNDARIES
43.              DO 30 K=2,N2
44.              H(1,K)=H(3,K)
45.              H(K,N2)=H(K,N)
46.    30        CONTINUE
47.    C    SWEEP INTERIOR NODES WITH FINITE DIFFERENCE FORM OF POISSON'S EQN.
48.              DO 40 I=2,N1
49.              KK=KBEG(I)
50.              DO 40 J=KK,N1
51.              R=0.
52.              OLDVAL=H(I,J)
53.              IF((I.EQ.2).AND.(J.EQ.N1))R=RWELL
54.              H(I,J)=(H(I-1,J)+H(I+1,J)+H(I,J-1)+H(I,J+1)+DX*DX*R/T)/4.
55.              H(I,J)=OMEGA*H(I,J) + (1.-OMEGA)*OLDVAL
56.              ERR=ABS(H(I,J)-OLDVAL)
57.              IF(ERR.GT.AMAX)AMAX=ERR
58.    40        CONTINUE
59.              IF(AMAX.GT.0.001)GO TO 35
60.    45        CONTINUE
61.              PRINT 50,NUMIT,((H(I,J),I=2,N2),J=1,N1)
62.    50        FORMAT(///1X,'NUMBER OF ITERATIONS IS',I4,///11(11F8.2//))
63.              STOP
64.              END
```

[49]

node in each column that is closest to 2000 m from the origin, the location of the well. The shaded area at the origin in Figure 3.4 is the top view of the small volume which represents the well. By Equation 3.8, $R(0, 0)$ is equal to -0.05 m day^{-1}. The recharge rate is zero for all other nodes representing similar small aquifer volumes. Lines 51 and 53 of the computer program ensure that the correct value of R is used in line 54, the finite difference form (Equation 3.7) of Poisson's equation.

The method of solution is by Gauss–Seidel iteration (line 54) with successive over relaxation (line 55). The relaxation parameter $OMEGA$ is chosen to be 1.8. The final accuracy of the iterative solution is sensitive to the choice of error tolerance, which is the maximum change in head at any node from one iteration to another. If the error tolerance is chosen too large, the program stops the iteration procedure too soon. In the example shown, the error tolerance was set equal to 0.001 (line 59).

The problem domain shown in Figure 3.4 is the upper right-hand quadrant of the problem illustrated in Figure 2.3. The analytical solution is given by the Thiem equation, which is a solution for radial flow to a pumping well under steady-state conditions. The Thiem equation is

$$h(r) - h(r_e) = \frac{Q}{2\pi T} \ln \frac{r}{r_e} \tag{3.9}$$

where, for this application, $r = \sqrt{x^2 + y^2}$, $Q = 2000$ m^3 day^{-1}, $T = 300$ m^2 day^{-1}, $r_e = 2000$ m, and $h(r_e) = 10$ m. If these values are substituted into Equation 3.9, the result is Equation 2.6. Equation 2.6 was used to compute the boundary heads for the region-near-a-well problem in Figure 2.3.

The computer solution of the drawdown problem is given in Figure 3.6. If these heads are compared with those of the analytical solution, we see very good agreement.

Figure 3.6
Output from the computer program in Figure 3.5. Heads are given in meters. The boxed area is the problem domain of the region-near-a-well example in Figures 2.3 and 2.8.

NUMBER OF ITERATIONS IS   63

```
10.00  10.00  10.00  10.00  10.00  10.00  10.00  10.00  10.00  10.00  10.00
 9.88   9.89   9.90   9.93  10.00  10.00  10.00  10.00  10.00  10.00  10.00
 9.76   9.76   9.79   9.83   9.88   9.93  10.00  10.00  10.00  10.00  10.00
 9.61   9.62   9.66   9.71   9.77   9.84   9.92  10.00  10.00  10.00  10.00
 9.45   9.46   9.51   9.57   9.65   9.74   9.83   9.92  10.00  10.00  10.00
 9.26   9.28   9.34   9.42   9.52   9.63   9.74   9.84   9.93  10.00  10.00
 9.02   9.05   9.14   9.26   9.39   9.53   9.65   9.77   9.88  10.00  10.00
 8.70   8.77   8.92   9.09   9.26   9.42   9.57   9.71   9.83   9.93  10.00
 8.26   8.41   8.67   8.92   9.14   9.34   9.51   9.66   9.79   9.90  10.00
 7.50  [7.96   8.41]  8.77   9.05   9.28   9.47   9.63   9.76   9.89  10.00
 5.84  [7.50   8.26]  8.71   9.02   9.26   9.45   9.61   9.76   9.88  10.00
```

3.5 UNCONFINED AQUIFER WITH DUPUIT ASSUMPTIONS

If an aquifer is unconfined, its saturated thickness varies with the water table height. We consider one-dimensional flow through an unconfined aquifer. The flow is one-dimensional under the Dupuit assumptions that (a) the flow is horizontal and (b) the hydraulic gradient is equal to the slope of the free surface. These assumptions imply the absence of a seepage face at the outflow boundary. To derive a governing equation for unconfined flow with Dupuit assumptions, we first do a continuity analysis for an infinitesimal slice of the aquifer (Figure 3.7). The slice is oriented perpendicular to the flow. The top of the repre-

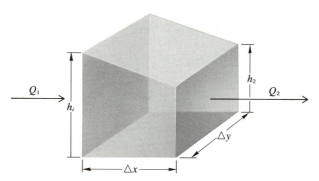

Figure 3.7
Infinitesimal volume of an unconfined aquifer. Top of the volume is the water table. Q_1 and Q_2 are volumetric rates of flow through this slice of the aquifer.

sentative volume is the water table and has height $h(x) = h_1$ at the left face and height $h(x) = h_2$ at the right face. The Dupuit assumption of horizontal flow means that the slope of the water table must be small. Let Q_1 and Q_2 be the *volumetric* rates of flow through the left and right faces, respectively. By applying Darcy's law and multiplying by the area of each face, we have

$$Q_2 - Q_1 = -K \, \Delta y \left[h_2 \frac{dh}{dx}\bigg|_{x_2} - h_1 \frac{dh}{dx}\bigg|_{x_1} \right] \tag{3.10}$$

If R is the recharge rate into the top of the representative volume, then, by continuity,

$$Q_2 - Q_1 = R \, \Delta x \, \Delta y \tag{3.11}$$

Each term within the brackets of Equation 3.10 can be expressed as a derivative of h^2 because $dh^2/dx = 2h(dh/dx)$. After dividing Equations 3.10 and 3.11 through by $\Delta x \, \Delta y$, we have

$$-\frac{K}{2} \left[\frac{\left.\dfrac{dh^2}{dx}\right|_{x_2} - \left.\dfrac{dh^2}{dx}\right|_{x_1}}{\Delta x} \right] = R \tag{3.12}$$

The term within the brackets in Equation 3.12 is d^2h^2/dx^2 in the limit as $\Delta x \to 0$. Thus, for one-dimensional, unconfined flow under the Dupuit assumptions,

$$\frac{K}{2} \frac{d^2h^2}{dx^2} = -R \tag{3.13}$$

For two-dimensional flow,

$$\frac{K}{2} \left(\frac{\partial^2 h^2}{\partial x^2} + \frac{\partial^2 h^2}{\partial y^2} \right) = -R \tag{3.14}$$

Equation 3.14 can be recognized to be Poisson's equation, Equation 3.2, if we make the change of variable $v = h^2$:

$$\frac{\partial^2 v}{\partial x^2} + \frac{\partial^2 v}{\partial y^2} = -\frac{2R}{K} \tag{3.15}$$

The boundary conditions must also be recast in terms of v instead of h. After a solution has been obtained in terms of v, the head is obtained by taking the square root of v.

Seepage Through a Dam

A classical problem of unconfined flow systems is locating the top boundary of the saturated zone in an earthen dam. In the cross section shown in Fig-

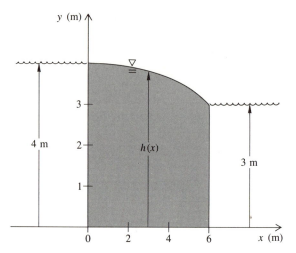

Figure 3.8
Schematic illustration of the water table and boundary
conditions for the example involving seepage through a
dam. The dam itself is not shown in the figure.

ure 3.8, the reservoir lake has an elevation of 4 m and the plunge pool has an
elevation of 3 m. The bottom of the dam at $y = 0$ rests on impermeable bedrock.
In this section, we solve the one-dimensional version of this problem using the
Dupuit approximation that flow is horizontal through the dam and that the
upper saturated boundary intersects the reservoir and plunge pool levels at
the two ends of the dam.

The exact formulation of this seepage problem requires that the top surface
be a no-flow boundary and that the head at each point on the top boundary
be equal to its elevation. In Chapter 6, we use the exact formulation and solve
the two-dimensional version of this problem with the finite element method.

In one dimension, the governing equation for this problem is Equation 3.13
with $R = 0$. The boundary conditions are that $h = 4$ m at $x = 0$ and that
$h = 3$ m at $x = 6$ m. The analytical solution can be derived by integration of
$d^2h^2/dx^2 = 0$ leading to the general solution $h^2 = a_1 x + a_2$, where a_1 and a_2
are constants determined by substitution of the boundary conditions. The
final analytical solution for the boundary conditions used in this problem is

$$h(x) = \sqrt{-1.17x + 16} \qquad (3.16)$$

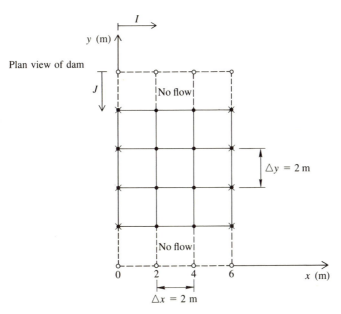

I

y (m)

Plan view of dam

J

No flow

$\triangle y = 2$ m

No flow

0   2   4   6 x (m)

$\triangle x = 2$ m

○ Fictitious node

✕ Specified head boundary node

Figure 3.9
Plan view of the finite difference grid used in the seepage through
a dam problem. The two no-flow boundaries parallel to the x
axis keep the flow one-dimensional in the x direction.

We now present a finite difference solution. Although the problem is one-dimensional, let us treat it in two-dimensional plan view (Figure 3.9) since this means that our existing programs need be modified only slightly. The boundaries at the top and bottom of Figure 3.9 are no-flow boundaries to force flow to be one-dimensional in the x direction. Remember that use of the Dupuit assumptions means that head does not vary vertically. The y dimension of the grid is arbitrary in length, but we have chosen it to be 6 m simply for the sake of keeping the spacing between nodes such that $\Delta x = \Delta y = 2$ m.

The computer program (Figure 3.10) has the same structure as earlier ones for Laplace's equation (for example, Figure 2.10), but the iteration formula is written in terms of v (line 30). We take the square root of v in line 38 to get back

Figure 3.10

Computer program to solve the one-dimensional version of the seepage through a dam problem under Dupuit assumptions.

```
1.      C    SEEPAGE THROUGH DAM UNDER DUPUIT ASSUMPTIONS
2.                DIMENSION V(11,11),H(11,11)
3.                TOL=0.01
4.                N=3
5.                N1=N+1
6.                N2=N+2
7.                N3=N+3
8.      C    SET CONSTANT HEAD BOUNDARY CONDITIONS IN TERMS OF V=H*H
9.      1         DO 5 J=1,N3
10.               V(1,J)=4*4
11.               V(N1,J)=3*3
12.     5         CONTINUE
13.     C    SET INITIAL GUESSES FOR REMAINING NODES
14.               DO 10 I=2,N
15.               DO 10 J=1,N3
16.               V(I,J)=4*4
17.     10        CONTINUE
18.               NUMIT=0
19.     35        AMAX=0.
20.               NUMIT=NUMIT+1
```

```
C    NO-FLOW BOUNDARY CONDITIONS PARALLEL TO X-AXIS
     DO 40 I=2,N
     V(I,1)=V(I,3)
     V(I,N3)=V(I,N1)
40   CONTINUE
C    ITERATE THROUGH INTERIOR NODES
     DO 50 J=2,N2
     DO 50 I=2,N
     OLDVAL=V(I,J)
     V(I,J)=(V(I-1,J)+V(I+1,J)+V(I,J-1)+V(I,J+1))/4.
     ERR=ABS(V(I,J)-OLDVAL)
     IF(ERR.GT.AMAX)AMAX=ERR
50   CONTINUE
     IF(AMAX.GT.TOL)GO TO 35
C    TAKE SQUARE ROOTS OF V TO GET HEADS
     DO 60 I=1,N1
     DO 60 J=1,N3
     H(I,J)=SQRT(V(I,J))
60   CONTINUE
     PRINT 70,NUMIT
70   FORMAT(1H1,'NUMBER OF ITERATIONS IS',I4)
     PRINT 75,((H(I,J),I=1,N1),J=2,N2)
75   FORMAT(1X,///4(4F8.2///))
     STOP
     END
```

21.
22.
23.
24.
25.
26.
27.
28.
29.
30.
31.
32.
33.
34.
35.
36.
37.
38.
39.
40.
41.
42.
43.
44.
45.

NUMBER OF ITERATIONS IS   12

4.00	3.70	3.37	3.00
4.00	3.70	3.37	3.00
4.00	3.70	3.37	3.00
4.00	3.70	3.37	3.00

Figure 3.11
Output from the computer program in Figure 3.10.
Configuration of nodes is the same as in Figure 3.9.
Heads are given in meters for an areal or plan view of
the dam.

to h. The solution is shown in Figure 3.11. The heads in each column of the solution are identical to the others, as they should be, since the top and bottom no-flow boundary conditions constrain flow to be one-dimensional. The values computed from the numerical solution for the height of the water table are $h = 3.70$ m at $x = 2$ m and $h = 3.37$ m at $x = 4$ m. These numbers agree to the first decimal place with those computed from the analytical solution, Equation 3.16.

Well Drawdown (Unconfined Aquifer)

We return to the well drawdown problem of Section 3.4, but we solve it in finite difference form for unconfined conditions with the Dupuit assumptions. The governing equation is Equation 3.14. We take the initial saturated thickness of the aquifer to be 10 m and refer heads to the base of the aquifer. The computer program is shown in Figure 3.12. It is very similar to the program for the confined aquifer case (Figure 3.5). Initial guesses for $H(I, J)$ are set to 5.0 m (line 15), and the boundary heads are set to 10.0 m (line 36). The change of variable $v = h^2$ is made in lines 38 to 41. Boundary conditions and Poisson's equation are cast in terms of v in lines 46 to 50 and lines 58 to 59, respectively.

Figure 3.12
Computer program to solve the drawdown example for an unconfined aquifer 10 m
thick using Dupuit assumptions.

```
1.     C    DRAWDOWN EXAMPLE - UNCONFINED AQUIFER
2.            DIMENSION H(12,12),V(12,12)
3.            DIMENSION KBEG(12)
4.            OMEGA=1.8
5.            N=10
6.            N1=N+1
7.            N2=N+2
8.     C    INPUT VARIABLES
9.            DX=200.
10.           RWELL=-2000./DX/DX
11.           COND=30.
12.    C    SET INITIAL GUESSES FOR H(I,J) REFERRED TO BASE OF AQUIFER
13.           DO 5 J=1,N2
14.           DO 5 I=1,N2
15.           H(I,J)=5.
16.    5      CONTINUE
17.    C    SET STATIC WATER LEVEL BOUNDARY CONDITION AT NODE IN EACH COLUMN
18.    C    WHICH IS CLOSEST TO BEING 10-KM FROM THE ORIGIN
19.           DO 25 I=2,N2
20.           DMIN=20000.
21.           DO 20 J=1,N1
22.           X=DX*(I-2)
23.           Y=2000.-DX*(J-1)
```

Figure 3.12 (Continued)

```
24.         DIST=SQRT(X*X+Y*Y)
25.         DELT=ABS(DIST-2000.)
26.         IF(DELT.GE.DMIN)GO TO 20
27.         DMIN=DELT
28.         KBEG(I)=J+1
29.   20    CONTINUE
30.   25    CONTINUE
31.   C  SET THE STATIC WATER LEVEL AT ALL NODES FARTHER
32.   C  THAN 2000-M OUT TO THE CORNERS, EQUAL TO 10 METERS
33.         DO 32 I=2,N2
34.         JBND=KBEG(I)-1
35.         DO 32 J=1,JBND
36.         H(I,J)=10.
37.   32    CONTINUE
38.         DO 33 I=1,N2
39.         DO 33 J=1,N2
40.         V(I,J)=H(I,J)*H(I,J)
41.   33    CONTINUE
42.         NUMIT=0
43.   35    AMAX=0.
44.         NUMIT=NUMIT+1
45.         IF(NUMIT.GT.500)GO TO 45
46.   C  LEFT AND BOTTOM SYMMETRY NO-FLOW BOUNDARIES
47.         DO 30 K=2,N2
48.         V(1,K)=V(3,K)
49.         V(K,N2)=V(K,N)
50.   30    CONTINUE
```

```
C       SWEEP INTERIOR NODES WITH FINITE DIFFERENCE FORM OF POISSON'S EQN.
        DO 40 I=2,N1
        KK=KBEG(I)
        DO 40 J=KK,N1
        R=0.
        OLDVAL=V(I,J)
        IF((I.EQ.2).AND.(J.EQ.N1))R=RWELL
        V(I,J)=(V(I-1,J)+V(I+1,J)+V(I,J-1)+V(I,J+1)+DX*DX*2.*R/COND)/4.
        V(I,J)=OMEGA*V(I,J)+(1.-OMEGA)*OLDVAL
        ERR=ABS(V(I,J)-OLDVAL)
        IF(ERR.GT.AMAX)AMAX=ERR
40      CONTINUE
        IF(AMAX.GT.0.001)GO TO 35
45      CONTINUE
        DO 48 I=1,N2
        DO 48 J=1,N2
        IF(V(I,J).LT.0.) V(I,J)=0.
        H(I,J)=SQRT(V(I,J))
48      CONTINUE
        PRINT 50,NUMIT,((H(I,J),I=2,N2),J=1,N1)
50      FORMAT(///1X,'NUMBER OF ITERATIONS IS',I4,///11(11F8.2//))
        STOP
        END
```

Finally, the solution is obtained in terms of $H(I, J)$ in line 68. If the potentiometric surface had been drawn down below the base of the aquifer, a negative value of v would have existed at the location of the well. In line 67, the head at the well is set equal to the base of the aquifer to handle that case. The computer printout of answers is shown in Figure 3.13. The answers are similar to those for the equivalent confined aquifer (Figure 3.6). The difference is caused by the fact that the product of hydraulic conductivity and saturated thickness is not a constant for the unconfined aquifer. For problems in which the drawdown is small relative to the thickness of the aquifer, the product Kh is nearly constant and equal to the transmissivity value of the equivalent confined aquifer.

The finite difference solution can also be compared with the analytical solution for an unconfined aquifer.

$$h^2(r) - h^2(r_e) = \frac{Q}{\pi K} \ln \frac{r}{r_e}$$ (3.17)

where $r = \sqrt{x^2 + y^2}$. In our numerical example, Q is 2000 m^3 day^{-1}, K is 30 m day^{-1}, r_e is 2000 m, and $h(r_e)$ is 10 m. The reference datum is the base of the aquifer. The comparison is good except for the node closest to the pumping well.

3.6   VALIDITY OF A NUMERICAL SOLUTION

In most of our examples, we checked the validity of the numerical solution by comparing the numbers generated using the numerical solution with those calculated from an analytical solution. However, analytical solutions are not available for many problems of practical interest. Numerical methods allow us to solve the governing equation in more than one dimension for complex boundary conditions and for heterogeneous and anisotropic aquifers, whereas most analytical solutions are restricted to consideration of homogeneous, isotropic aquifers. An important point which is sometimes overlooked is the necessity to verify the validity of every numerical solution. Therefore, several checks on the correctness of the solution should be made routinely when an analytical solution is not available. Specifically, it is a good practice to check the sensitivity of the solution to the choice of error tolerance and nodal spacing and to perform a mass balance calculation. Each of these validity checks is discussed here.

In the well-drawdown problem discussed in Section 3.4, we pointed out that the solution was sensitive to the value chosen for the error tolerance. If the

Figure 3.13
Output from the computer program in Figure 3.12. Heads are in meters above the base of the aquifer.

NUMBER OF ITERATIONS IS 89

10.00	10.00	10.00	10.00	10.00	10.00	10.00	10.00	10.00	10.00	10.00
10.00	10.00	10.00	10.00	10.00	10.00	10.00	10.00	10.00	10.00	10.00
10.00	10.00	10.00	10.00	10.00	9.93	9.88	9.83	9.78	9.76	9.75
10.00	10.00	10.00	10.00	9.92	9.84	9.77	9.70	9.65	9.62	9.61
10.00	10.00	10.00	9.92	9.82	9.73	9.64	9.56	9.50	9.45	9.43
10.00	10.00	9.93	9.84	9.73	9.62	9.51	9.41	9.31	9.25	9.23
10.00	10.00	9.88	9.77	9.64	9.51	9.37	9.23	9.10	9.00	8.96
10.00	9.93	9.83	9.70	9.56	9.41	9.23	9.04	8.85	8.69	8.61
10.00	9.90	9.78	9.65	9.50	9.31	9.10	8.85	8.56	8.26	8.07
10.00	9.89	9.76	9.62	9.45	9.25	9.00	8.69	8.26	7.69	7.08
10.00	9.88	9.75	9.61	9.43	9.23	8.96	8.61	8.07	7.08	4.09

error tolerance was set too high, the solution did not match the analytical solution. As a general rule, several computer runs should be made using successively smaller values for the error tolerance until a small enough value is selected and the solution does not change within the desired range of accuracy. The solution might also be improved through the use of smaller nodal spacing. Several computer runs could be made with successively smaller nodal spacing until the solution does not change within the desired range of accuracy. However, it is generally a time-consuming process to modify the finite difference grid and redefine the input. Instead, most modelers rely on the third type of validity check—the mass balance calculation.

A mass balance calculation is an expression of the fact that, at steady state, the amount of water entering the system equals the amount leaving the system. If inflow does not equal outflow for a steady-state computer simulation, there may be something wrong with the numerical solution. However, a large error in the mass balance may also indicate a lack of precision in the mass balance calculation itself.

The goal of most computer simulations is to predict the effects of some proposed management scheme on a particular groundwater system. The final test of a numerical model is to determine whether it successfully simulates field observations. Such a model is said to be calibrated and verified. The process of calibration and verification generally requires adjustments of model parameters and/or boundary conditions, and this process is discussed more fully in Chapter 5.

Notes and Additional Reading

1. The Dupuit assumptions are discussed in most textbooks on groundwater hydrology. For example, see pp. 188–189 of Freeze and Cherry (1979) or pp. 132–139 of Fetter (1980). Remson et al. (1971, pp. 51–54) also discuss the use of the Dupuit assumptions.

2. The validity of the Dupuit assumptions has been under scrutiny in the literature for a number of years. The calculations based on the use of the Dupuit assumptions are acceptable when the slope of the water table is small and when the depth of the unconfined flow field is shallow. For more information on the limitations of the Dupuit assumptions, see Kirkham (1967), Childs and Young (1968), Murray and Monkmeyer (1973), and Verma and Brutsaert (1971).

3. For a discussion of the problem of selecting nodal spacings which are sufficiently small, see Remson et al. (1971, pp. 150–152) and Rushton and Tomlinson (1977), whose results are also summarized by Rushton and Redshaw (1979, pp. 184–185).

Problems

3.1 Refer to the Long Island example discussed in Section 3.3 and find the analytical solution for this problem using

$$\frac{d^2h}{dx^2} = -\frac{R}{T}$$

subject to the boundary conditions

$$h(-\ell) = 0$$
$$h(+\ell) = 0$$

Compare the result with Equation 3.5.

3.2 Modify the computer program in either Figure 2.7 or Figure 2.10 to solve for the recharge rate R as well as the distribution of heads throughout the island in Figure 3.3, where $T = 10,000 \text{ ft}^2 \text{ day}^{-1}$, $\ell = 12,000 \text{ ft}$, and the elevation of the water level in the well at the center of the island is 20 ft above sea level. The analytical solution to this problem is translated from the equivalent heat conduction problem (Carslaw and Jaeger, 1959, p. 171) to be

$$h(x, y) = \frac{R(a^2 - x^2)}{2T} - \frac{16Ra^2}{T\pi^3} \sum_{n=0}^{\infty} \frac{(-1)^n \cos \dfrac{(2n + 1)\pi x}{2a} \cosh \dfrac{(2n + 1)\pi y}{2a}}{(2n + 1)^3 \cosh \dfrac{(2n + 1)\pi b}{2a}}$$

where $a = \ell$ and $b = 2\ell$.

3.3 Modify the program in Figure 3.5 so that the grid interval $\dot{D}X$ is equal to 100 m instead of 200 m and so that four nodes represent the well. Compare the results with those shown in Figure 3.6.

3.4 Consider the island recharge problem discussed in Section 3.4 and Problem 3.2. Suppose the aquifer is 100 ft thick at the shoreline (Figure 3.14). Solve the island

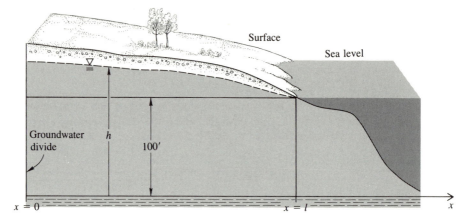

Figure 3.14
Cross section of the island recharge problem for unconfined conditions.

recharge problem for unconfined flow with Dupuit assumptions in one dimension analytically and in two dimensions by finite differences. Refer to Section 3.5, Equations 3.13 and 3.14. For both cases, formulate the appropriate mathematical model in terms of governing equation and boundary conditions.

3.5 Check the validity of your solution to the island recharge example in Problem 3.2 by changing the error tolerance, reducing the nodal spacing, and performing a mass balance computation. To perform the mass balance computation, you must add an additional section to your program to compute total inflow and outflow. The inflow is the volume of water entering the aquifer from areal recharge. This value should equal the volume of water leaving the aquifer by discharge to the ocean. The discharge to the ocean can be determined by performing a series of Darcy's law computations along the outflow boundaries.

CHAPTER 4

Finite Differences: Transient Flow

4.1 TRANSIENT FLOW EQUATION

A transient problem is one in which the unknown variable is time dependent. That is, heads change with time in a transient problem. Transient problems are also called time dependent, unsteady, nonequilibrium, or nonsteady-state problems. In the derivation of the governing equation for transient conditions, the continuity equation is modified such that the volume outflow rate equals the volume inflow rate plus the rate of release of water from storage. Therefore, we must introduce an expression for the rate of release of water from storage.

This expression involves the use of the storage coefficient S, which represents the volume of water released from storage per unit area of aquifer per unit decline in head. That is,

$$S = \frac{-\Delta V_w}{\Delta x\, \Delta y\, \Delta h} \qquad (4.1)$$

where ΔV_w is the volume of water released from storage within the elemental volume whose area is $\Delta x\, \Delta y$ and whose thickness is b. The presence of the minus sign in Equation 4.1 is explained by the following sign convention. When water is released from storage, ΔV_w is positive and Δh is negative; when water is taken up into storage, ΔV_w is negative and Δh is positive. The rate of release from storage is $\Delta V_w/\Delta t$ and can be written as $-S\,\Delta x\, \Delta y(\Delta h/\Delta t)$. As $\Delta t \to 0$, this expression becomes $-S(\partial h/\partial t)(\Delta x\, \Delta y)$. Therefore, the form of the continuity equation, Equation 3.1, for transient conditions is

$$\frac{\partial q_x}{\partial x}\Delta x(b\, \Delta y) + \frac{\partial q_y}{\partial y}\Delta y(b\, \Delta x) = R(x, y, t)\Delta x\, \Delta y - S\frac{\partial h}{\partial t}(\Delta x\, \Delta y) \qquad (4.2)$$

Substituting Darcy's law for q_x and q_y and dividing through by $-T\,\Delta x\, \Delta y$, where $T = Kb$, yields the transient flow equation

$$\frac{\partial^2 h}{\partial x^2} + \frac{\partial^2 h}{\partial y^2} = \frac{S}{T}\frac{\partial h}{\partial t} - \frac{R(x, y, t)}{T} \qquad (4.3)$$

In this chapter, we demonstrate how this equation is solved numerically using either an explicit or implicit approximation. In the next chapter, we consider solution techniques in which matrix methods are used directly, or in combination with iteration, to solve the set of algebraic equations generated using finite difference approximations.

We saw in Chapter 1 that, in order to solve Laplace's equation, it is necessary to prescribe boundary conditions. A governing equation, such as Laplace's equation or Poisson's equation, and boundary conditions constitute a mathematical model of a groundwater flow system at steady state. For nonsteady-state conditions, we must not only know the boundary conditions but also have a meaningful set of head values at the start of the simulation. These head values are the initial conditions. We can then step the solution forward in time starting from the initial conditions and solve for heads at several different time levels. When heads no longer change with time, we have reached steady state.

4.2 EXPLICIT FINITE DIFFERENCE APPROXIMATION

Suppose we have already established a grid of nodal points in the problem domain. We know how to write finite difference approximations for space derivatives. The head h_{ij} at each nodal point now is also a function of time. We need a finite difference approximation for the time derivative $\partial h/\partial t$ at each nodal point. Just as with the space domain, we divide the time domain into discrete points. For time level (n) where $(n + 1)$ represents one time step later and $(n - 1)$ represents one time step earlier than time (n), we could set

$$\frac{\partial h}{\partial t} \simeq \frac{h_{ij}^{n+1} - h_{ij}^{n}}{\Delta t} \tag{4.4}$$

The superscript is the time index and Δt is the length of the time step. Equation 4.4 is called a forward difference approximation relative to time (n). The equation

$$\frac{\partial h}{\partial t} \simeq \frac{h_{ij}^{n} - h_{ij}^{n-1}}{\Delta t} \tag{4.5}$$

is called a backward difference approximation relative to time (n). Equation 4.4 could also be considered a backward difference relative to time $(n + 1)$. The use of the central difference approximation

$$\frac{\partial h}{\partial t} \simeq \frac{h_{ij}^{n+1} - h_{ij}^{n-1}}{2\Delta t} \tag{4.6}$$

should be avoided because this approximation is unconditionally unstable (Remson et al., 1971).

We use the forward difference approximation and write the finite difference form of the transient flow equation, Equation 4.3.

$$\frac{h_{i+1, j}^{n} - 2h_{ij}^{n} + h_{i-1, j}^{n}}{(\Delta x)^2} + \frac{h_{i, j+1}^{n} - 2h_{ij}^{n} + h_{i, j-1}^{n}}{(\Delta y)^2} = \frac{S}{T} \frac{h_{ij}^{n+1} - h_{ij}^{n}}{\Delta t} - \frac{R_{ij}^{n}}{T} \tag{4.7}$$

Note that the space derivatives have been evaluated at the known time step (n).

If we let $\Delta x = \Delta y = a$ and solve for h_{ij}^{n+1}, then

$$
\begin{aligned}
h_{ij}^{n+1} = \left(1 - \frac{4T\,\Delta t}{Sa^2}\right)h_{ij}^n \\
+ \left(\frac{4T\,\Delta t}{Sa^2}\right)\left(\frac{h_{i+1,j}^n + h_{i-1,j}^n + h_{i,j+1}^n + h_{i,j-1}^n}{4}\right) + \frac{R_{ij}^n\,\Delta t}{S}
\end{aligned} \quad (4.8)
$$

Equation 4.8 is called an explicit or forward difference approximation because h_{ij}^{n+1} is evaluated in terms of the known or old values of h at the nodes surrounding (i, j). We can proceed in a recursive manner of computing heads over the spatial domain for future times after we have been given the heads at some initial time. Notice that if $R_{ij}^n = 0$, Equation 4.8 is essentially the successive over relaxation formula given by Equation 2.13, where $\omega = 4T\,\Delta t/Sa^2$.

Validity of the Explicit Solution

In Section 3.6, we mentioned that the head values generated using a numerical solution may be sensitive to nodal spacing. For transient problems, the head values are also dependent on the choice of Δt. A finite difference approximation is said to converge if the finite difference solution approaches the correct solution as the nodal spacing and Δt approach zero. The finite difference approximation is stable if, as the solution marches forward in time, the errors are not amplified such that the solution becomes invalid. For an explicit approximation, it can be demonstrated that the ratio $T\,\Delta t/Sa^2$ in Equation 4.8 must be kept sufficiently small for the solution to be stable. In the one-dimensional case where flow occurs only in the x direction, the parameter $T\,\Delta t/S(\Delta x)^2$ must be less than or equal to 0.5 (Remson et al., 1971). For the two-dimensional case where $\Delta x = \Delta y = a$, $T\,\Delta t/Sa^2$ must be less than or equal to 0.25 (Rushton and Redshaw, 1979). If this parameter is not sufficiently small, errors will be amplified as the solution progresses, and the solution is said to be unstable.

Aquifer Response to Sudden Change in Reservoir Level

Consider the confined aquifer shown in Figure 4.1. Initially head is equal to 16 m everywhere in the aquifer. We wish to simulate changes in head through time if, at $t = 0$, we suddenly drop the water level in the reservoir at $x = \ell$ from 16 m to 11 m. The aquifer parameters are $T = 0.02$ m^2 minute^{-1} and $S =$

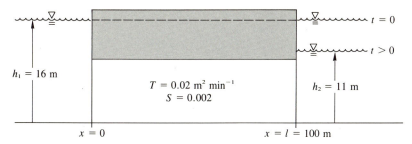

Figure 4.1
The reservoir example. Boundary and initial conditions are shown.
Flow is one-dimensional in the x direction.

0.002. If we can neglect flow in the y direction, we can use the one-dimensional governing equation

$$\frac{\partial^2 h}{\partial x^2} = \frac{S}{T}\frac{\partial h}{\partial t} \tag{4.9}$$

The boundary conditions are $h(0, t) = h_1$ and $h(\ell, t) = h_2$ for $t > 0$, where $h_1 = 16$ m, $h_2 = 11$ m, and $\ell = 100$ m. The initial condition is $h(x, 0) = h_1$ for $0 \le x \le \ell$.

The explicit finite difference approximation for Equation 4.9 is

$$\frac{h_{i+1}^n - 2h_i^n + h_{i-1}^n}{(\Delta x)^2} = \frac{S}{T}\left(\frac{h_i^{n+1} - h_i^n}{\Delta t}\right) \tag{4.10}$$

Solving for h_i^{n+1}, we obtain

$$h_i^{n+1} = h_i^n + \frac{T\,\Delta t}{S}\left(\frac{h_{i+1}^n - 2h_i^n + h_{i-1}^n}{(\Delta x)^2}\right) \tag{4.11}$$

A computer program which solves this model using Equation 4.11 is given in Figure 4.2. The nodal spacing is chosen to be $\Delta x = 10$ m. For the explicit scheme to be stable, $T\,\Delta t/S(\Delta x)^2$ must be less than 0.5. That is, for this problem, Δt must be less than or equal to 5 minutes. Using Δt equal to 5 minutes, the results are shown in Figure 4.3 for every other time step. The profile of the

Figure 4.2

Computer program for the reservoir example using an explicit finite difference solution technique.

```
1.    C    RESERVOIR EXAMPLE
2.              DIMENSION HOLD(25),HNEW(25)
3.    C    HOLD IS HEAD AT THE TIME STEP N
4.    C    HNEW IS HEAD AT THE TIME STEP N+1
5.    C    DEFINE PARAMETERS
6.              DT=5.
7.              DX=10.
8.              T=0.02
9.              S=0.002
10.             NX=11
11.             NLX=NX-1
12.   C    DEFINE INITIAL AND BOUNDARY CONDITIONS
13.   C    HO IS THE INITIAL HEAD
14.             HO=16.
15.             DO 10 I=1,NX
16.             HOLD(I)=HO
17.             HNEW(I)=HO
18.      10 CONTINUE
19.   C    CHANGE HEAD AT BOUNDARY FOR TIME GREATER THAN ZERO
20.             HOLD(NX)=11.
```

```
      C COMPUTE HEADS THROUGH TIME
            PRINT 120
        120 FORMAT(1H1,31X,'HEAD',36X,'TIME',//)
            KOUNT=1
            KPRINT=2
            TIME=DT
            NEND=100.
            DO 50 N=1,NEND
            DO 20 I=2,NLX
            F1=DT*T/S
            D2H=(HOLD(I+1)-2.*HOLD(I)+HOLD(I-1))/(DX*DX)
            HNEW(I)=HOLD(I)+(F1*D2H)
         20 CONTINUE
            DO 30 I=1,NLX
            HOLD(I)=HNEW(I)
         30 CONTINUE
      C PRINT RESULTS FOR EVERY OTHER TIME STEP
            IF(KOUNT.NE.KPRINT) GO TO 49
            WRITE(6,40) (HOLD(I),I=1,NX),TIME
            KOUNT=0
         49 TIME=TIME+DT
            KOUNT=KOUNT+1
         50 CONTINUE
         40 FORMAT(1X,11F6.2,1F10.2)
            STOP
            END
```

Figure 4.3
Output from the computer program in Figure 4.2.

TIME	HEAD										
10.00	11.00	13.50	14.75	16.00	16.00	16.00	16.00	16.00	16.00	16.00	16.00
20.00	11.00	12.88	14.13	15.38	15.69	15.84	15.92	15.96	15.98	15.99	16.00
30.00	11.00	12.56	13.73	14.91	15.38	15.65	15.80	15.89	15.94	15.97	16.00
40.00	11.00	12.37	13.46	14.55	15.10	15.45	15.67	15.81	15.89	15.94	16.00
50.00	11.00	12.23	13.26	14.28	14.87	15.27	15.54	15.71	15.83	15.91	16.00
60.00	11.00	12.13	13.09	14.06	14.67	15.10	15.41	15.62	15.77	15.88	16.00
70.00	11.00	12.05	12.96	13.88	14.49	14.95	15.28	15.53	15.70	15.85	16.00
80.00	11.00	11.98	12.85	13.73	14.34	14.81	15.17	15.44	15.64	15.82	16.00
90.00	11.00	11.93	12.76	13.60	14.20	14.69	15.06	15.35	15.59	15.79	16.00
100.00	11.00	11.88	12.68	13.48	14.08	14.57	14.96	15.28	15.53	15.77	16.00
110.00	11.00	11.84	12.61	13.38	13.98	14.47	14.87	15.20	15.49	15.74	16.00
120.00	11.00	11.81	12.55	13.30	13.88	14.38	14.79	15.14	15.44	15.72	16.00
130.00	11.00	11.78	12.50	13.22	13.80	14.29	14.72	15.08	15.40	15.70	16.00
140.00	11.00	11.75	12.45	13.15	13.72	14.22	14.65	15.02	15.36	15.68	16.00
150.00	11.00	11.72	12.40	13.08	13.65	14.15	14.59	14.97	15.33	15.66	16.00
160.00	11.00	11.70	12.36	13.03	13.59	14.09	14.53	14.93	15.30	15.65	16.00
170.00	11.00	11.68	12.33	12.98	13.53	14.03	14.48	14.89	15.27	15.63	16.00
180.00	11.00	11.66	12.30	12.93	13.48	13.98	14.43	14.85	15.24	15.62	16.00
190.00	11.00	11.65	12.27	12.89	13.44	13.93	14.39	14.82	15.22	15.61	16.00
200.00	11.00	11.63	12.24	12.85	13.39	13.89	14.36	14.79	15.20	15.60	16.00
210.00	11.00	11.62	12.22	12.82	13.36	13.86	14.32	14.76	15.18	15.60	16.00
220.00	11.00	11.61	12.20	12.79	13.32	13.82	14.29	14.74	15.18	15.59	16.00
230.00	11.00	11.60	12.18	12.76	13.29	13.80	14.27	14.76	15.18	15.60	16.00
240.00	11.00	11.59	12.16	12.74	13.26	13.79	14.26	14.74	15.16	15.59	16.00

250.00	11.00	11.58	12.15	12.71	13.24	13.76	14.24	14.71	15.15	15.58	16.00
260.00	11.00	11.57	12.13	12.69	13.22	13.74	14.22	14.69	15.13	15.57	16.00
270.00	11.00	11.57	12.12	12.67	13.19	13.72	14.19	14.67	15.12	15.57	16.00
280.00	11.00	11.56	12.11	12.66	13.18	13.69	14.18	14.66	15.11	15.56	16.00
290.00	11.00	11.55	12.10	12.64	13.16	13.68	14.16	14.64	15.10	15.55	16.00
300.00	11.00	11.55	12.09	12.63	13.14	13.66	14.14	14.63	15.09	15.55	16.00
310.00	11.00	11.54	12.08	12.61	13.13	13.64	14.13	14.62	15.08	15.54	16.00
320.00	11.00	11.54	12.07	12.60	13.12	13.63	14.12	14.61	15.07	15.54	16.00
330.00	11.00	11.54	12.07	12.59	13.11	13.62	14.11	14.60	15.07	15.53	16.00
340.00	11.00	11.53	12.06	12.58	13.10	13.61	14.10	14.59	15.06	15.53	16.00
350.00	11.00	11.53	12.05	12.57	13.09	13.60	14.09	14.58	15.05	15.52	16.00
360.00	11.00	11.53	12.05	12.56	13.08	13.59	14.08	14.57	15.05	15.52	16.00
370.00	11.00	11.52	12.04	12.56	13.07	13.58	14.07	14.56	15.04	15.52	16.00
380.00	11.00	11.52	12.04	12.55	13.06	13.57	14.06	14.56	15.04	15.52	16.00
390.00	11.00	11.52	12.03	12.55	13.05	13.56	14.06	14.55	15.04	15.52	16.00
400.00	11.00	11.52	12.03	12.54	13.05	13.56	14.05	14.55	15.03	15.51	16.00
410.00	11.00	11.51	12.03	12.54	13.04	13.55	14.05	14.54	15.03	15.51	16.00
420.00	11.00	11.51	12.02	12.53	13.04	13.54	14.04	14.54	15.03	15.51	16.00
430.00	11.00	11.51	12.02	12.53	13.03	13.54	14.04	14.53	15.02	15.51	16.00
440.00	11.00	11.51	12.02	12.53	13.03	13.54	14.04	14.53	15.02	15.51	16.00
450.00	11.00	11.51	12.02	12.52	13.03	13.53	14.03	14.53	15.02	15.51	16.00
460.00	11.00	11.51	12.01	12.52	13.02	13.53	14.03	14.52	15.02	15.51	16.00
470.00	11.00	11.51	12.01	12.52	13.02	13.53	14.03	14.52	15.02	15.51	16.00
480.00	11.00	11.51	12.01	12.52	13.02	13.52	14.02	14.52	15.01	15.51	16.00
490.00	11.00	11.51	12.01	12.52	13.02	13.52	14.02	14.52	15.01	15.51	16.00
500.00	11.00	11.51	12.01	12.52	13.02	13.52	14.02	14.52	15.01	15.51	16.00

[75]

potentiometric surface is graphed for several times in Figure 4.4a. Figure 4.4b demonstrates that, if $\Delta t = 8$ minutes, the explicit scheme becomes unstable.

Notice in Figure 4.4a that the numerical solution for $t = 400$ minutes is essentially equal to the analytical steady-state solution of Laplace's equation subject to the boundary conditions $h(0) = h_1$ and $h(\ell) = h_2$. Readers should verify that the steady-state analytical solution is

$$h(x) = \left(\frac{h_2 - h_1}{\ell}\right)x + h_1 \tag{4.12}$$

4.3   IMPLICIT FINITE DIFFERENCE APPROXIMATION

In Equation 4.7, the finite difference approximation of the space derivatives $\partial^2 h/\partial x^2$ and $\partial^2 h/\partial y^2$ is made at the time level (n). Because h_{ij}^{n+1} depends only on values of the head at the previous time step, we obtained an explicit formula, Equation 4.8, for h_{ij}^{n+1}. Between time step (n) and time step $(n + 1)$, the heads at all the nodes are changing, and the use of head values at the time level (n) to approximate the space derivatives is valid only if we take relatively small time steps. We can improve the approximation by evaluating the space derivatives somewhere between $t = n\,\Delta t$ and $t = (n + 1)\,\Delta t$. We do this by using a weighted average of the approximations at (n) and $(n + 1)$. The weighting parameter is represented by α, and it lies between 0 and 1. If time step $(n + 1)$ is weighted by α and time step (n) is weighted by $(1 - \alpha)$, then

$$\frac{\partial^2 h}{\partial x^2} \simeq \alpha\,\frac{h_{i+1,j}^{n+1} - 2h_{ij}^{n+1} + h_{i-1,j}^{n+1}}{(\Delta x)^2} + (1 - \alpha)\,\frac{h_{i+1,j}^{n} - 2h_{ij}^{n} + h_{i-1,j}^{n}}{(\Delta x)^2} \tag{4.13}$$

A similar expression could be written for $\partial^2 h/\partial y^2$. To shorten our expressions, we again adopt notation for the average head at the four points surrounding node (i, j).

$$\tilde{h}_{ij}^{n} = \frac{h_{i-1,j}^{n} + h_{i+1,j}^{n} + h_{i,j-1}^{n} + h_{i,j+1}^{n}}{4} \tag{4.14}$$

The transient flow equation, Equation 4.3, can then be approximated by

$$\alpha(\tilde{h}_{ij}^{n+1} - h_{ij}^{n+1}) + (1 - \alpha)(\tilde{h}_{ij}^{n} - h_{ij}^{n}) = \frac{a^2 S}{4T}\,\frac{h_{ij}^{n+1} - h_{ij}^{n}}{\Delta t} - \frac{a^2 R_{ij}^{n}}{4T} \tag{4.15}$$

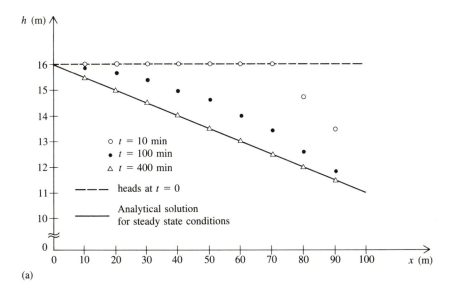

(a)

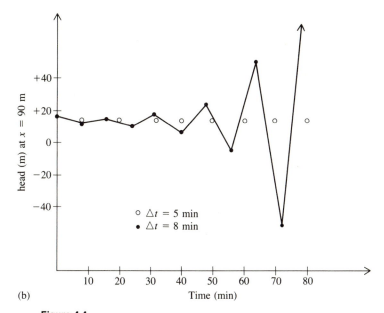

(b)

Figure 4.4
Graphs illustrating output from reservoir example.
(a) Change in head with distance for $t = 10$ min, $t = 100$ min, and $t = 400$ min. Initial heads (that is, $t = 0$) and the analytical solution for steady-state conditions are also shown.
(b) Change in head with time at $x = 90$ m when $\Delta t = 5$ min (solution is stable) and $\Delta t = 8$ min (solution is unstable).

where $\Delta x = \Delta y = a$. If $\alpha = 0$, Equation 4.15 reduces to the explicit formula, Equation 4.8. In general, however, Equation 4.15 contains as unknowns the heads at the four nodes surrounding (i, j) at the advanced time level $(n + 1)$ in addition to the head at point (i, j). Equation 4.15 is called an implicit approximation because h_{ij}^{n+1} is not expressed explicity in terms of known quantities.

The implicit approximation (also called a backward difference formulation) requires that heads at the $(n + 1)$ time level be obtained as the solution of a system of linear equations. In Equation 4.15, we put all the unknowns on the left-hand side and all the knowns on the right-hand side and multiply each side by negative 1.

$$\left(\frac{a^2 S}{4T \, \Delta t} + \alpha \right) h_{ij}^{n+1} - \alpha \tilde{h}_{ij}^{n+1} = \frac{a^2 S}{4T \, \Delta t} h_{ij}^n + (1 - \alpha)(\tilde{h}_{ij}^n - h_{ij}^n) + \frac{a^2 R_{ij}^n}{4T} \quad (4.16)$$

Equation 4.16 represents a system of linear equations for h_{ij} at time level $(n + 1)$. This system could be solved directly using matrix methods or by iteration in a manner similar to that used to solve the system of linear equations for the steady-state Laplace's or Poisson's equations. The difference is that, for the transient flow problem, a new system of equations must be solved at each time step. In this section, we consider the solution of Equation 4.16 by iteration. In Chapter 5, we consider the solution of Equation 4.16 by matrix methods.

The Gauss–Seidel iteration equation is obtained by solving Equation 4.16 for h_{ij}^{n+1}.

$$h_{ij}^{n+1} = \frac{1}{[(a^2 S/4T \, \Delta t) + \alpha]} \left[\alpha \tilde{h}_{ij}^{n+1} + \frac{a^2 S}{4T \, \Delta t} h_{ij}^n + (1 - \alpha)(\tilde{h}_{ij}^n - h_{ij}^n) + \frac{a^2 R_{ij}^n}{4T} \right] \quad (4.17)$$

Keep in mind that we must solve a new system of equations for each time step. If, for example, a simulation is to be run for 500 days and $\Delta t = 5$ days, then it is necessary to solve 100 systems of linear equations, whereas, for the steady-state problem, only one system of equations must be solved.

The parameter α is selected by the modeler. For $\alpha = 1$, the space derivatives are approximated solely at the advanced time level $(n + 1)$, and the finite difference scheme is said to be fully implicit. Use of the fully implicit scheme implies that we presume that the value of the space derivatives at the future time is the best approximation. Use of the explicit scheme $\alpha = 0$ implies that the value of the space derivatives at the old time level is the best approximation. If we select $\alpha = \frac{1}{2}$, we presume that the best value lies halfway between time levels (n) and $(n + 1)$. The finite difference approximation associated with a choice of $\alpha = \frac{1}{2}$ is called the Crank–Nicolson method.

Well Drawdown (Theis Problem)

Let us consider a transient well drawdown problem related to the one discussed in Section 3.4. A well is discharging at a constant rate Q from an areally extensive confined aquifer where the potentiometric surface is initially horizontal and equal to h_0. This problem is the one solved by Theis (1935). The drawdown at a radius r from the well is

$$h_0 - h = \frac{Q}{4\pi T} W(u) \qquad (4.18)$$

where

$$W(u) = \int_u^\infty \frac{e^{-\psi}}{\psi} d\psi \qquad (4.19)$$

and

$$u = \frac{r^2 S}{4Tt} \qquad (4.20)$$

$W(u)$ is called the well function and can be found in tabular form in most introductory texts in hydrogeology or well hydraulics (for example, Freeze and Cherry, 1979, p. 318, or Fetter, 1980, p. 460).

Although we have an analytical solution, we use this problem to illustrate the implicit finite difference technique. The mesh-centered grid chosen for this problem is shown in Figure 4.5, where $\Delta x = \Delta y = 100$ m. The left and lower boundaries are approximated as no-flow (or symmetry) boundaries so that we can deal with only one-fourth of the aquifer. In the computer model, the right and upper boundaries are also taken to be no-flow boundaries and are arbitrarily located 2000 m from the well. As we shall see, the numerical solution deviates from the Theis solution when the cone of depression extends to the upper and right boundaries.

The aquifer coefficients are $T = 300$ m^2 day^{-1} and $S = 0.002$. The well is discharging at a constant rate of 2000 m^3 day^{-1}. The area of influence of the well node is the shaded area shown in Figure 4.5. This problem is solved

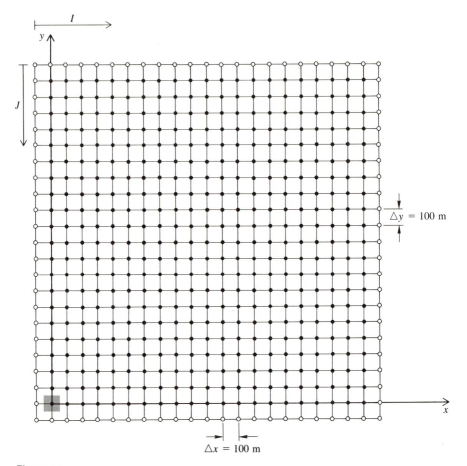

Figure 4.5
Finite difference grid for the transient version of the well drawdown example. Shaded cell represents the pumping well. Nodes represented by open circles are fictitious nodes used to simulate no-flow boundaries.

by using the computer program given in Figure 4.6. Note that, in line 11, a value of *ALPHA* is selected by the modeler. Small time steps are taken at the beginning of the simulation when heads are changing rapidly. An initial Δt of 0.01 day was used (line 9). In line 74, Δt is increased after every time step so that larger time steps are taken as the rate of change of head decreases with time. Output for $t = 13.1$ days is shown in Figure 4.7. Also, Table 4.1 shows

Table 4.1
Drawdowns $(h_0 - h)$ in meters at $r = 100$ m from a well discharging from a confined aquifer[†]

t (days)	Δt (days)	$(h_0 - h)$ for $\alpha = 0$	$(h_0 - h)$ for $\alpha = \frac{1}{2}$	Number of iterations	$(h_0 - h)$ for $\alpha = 1$	Number of iterations	$(h_0 - h)$ for Theis solution
0.01	0.01	0.00	0.05	4	0.06	4	0.04
0.05	0.02	0.46	0.41	5	0.38	6	0.42
0.13	0.05	0.83	0.88	7	0.82	9	0.86
0.49	0.17	0.22×10^3 *	1.55	14	1.49	19	1.51
1.13	0.38	0.27×10^6 *	1.98	22	1.92	31	1.94
5.82	1.95	0.60×10^{14} *	2.81	59	2.71	74	2.80
13.12	4.38	0.10×10^{20} *	3.26	101	3.10	114	3.23

* Solution unstable
[†] Error tolerance = 0.001

the results for a distance of 100 m from the well, that is, the two nodes nearest the origin. Drawdowns are given at several time levels for the explicit ($\alpha = 0$), Crank–Nicolson ($\alpha = 0.5$), and fully implicit ($\alpha = 1$) methods. The numerical values are compared with the answers calculated from the analytical solution of Theis, Equation 4.18, for $r = 100$ m. The explicit scheme becomes unstable when Δt is greater than 0.05 days, whereas the Crank–Nicolson and fully implicit methods are stable for any value of Δt. Note that the implicit scheme requires a greater number of iterations than the Crank–Nicolson method and that a greater number of iterations is required as time steps become larger for both methods. If we had not increased the size of the time step as rapidly, fewer iterations would have been required between time steps. Some of the error in the numerical results is caused by choice of error tolerance and at the later

Figure 4.6
Finite difference computer program for solving the transient flow equation for the well
drawdown example for a confined aquifer.

```
1.    C     DRAWDOWN EXAMPLE - CONFINED AQUIFER - TRANSIENT CONDITIONS
2.              DIMENSION HNEW(23,23),HOLD(23,23),R(23,23),DD(23,23)
3.    C     DEFINE INPUT PARAMETERS
4.              S=0.002
5.              T=300.
6.    C     HO IS INITIAL HEAD
7.              HO=10.
8.              DX=100.
9.              DT=0.01
10.   C     USE CRANK-NICOLSON APPROXIMATION
11.             ALPHA=0.5
12.   C     SET ERROR TOLERANCE
13.             TOL=0.001
14.   C     INITIALIZE ARRAYS
15.   C     HOLD IS HEAD AT THE TIME STEP N
16.   C     HNEW IS HEAD AT THE TIME STEP N+1
17.             DO 4 I=1,23
18.             DO 4 J=1,23
19.             HNEW(I,J)=HO
20.             HOLD(I,J)=HO
21.             R(I,J)=0.
22.         4 CONTINUE
23.   C     DEFINE PUMPING RATE AS RECHARGE TO CELL AT ORIGIN
24.             R(2,22)=-2000./DX/DX
25.             TIME=0.
```

```
C     START TIME STEPS
C     AT EACH TIME STEP SOLVE SYSTEM OF EQUATIONS BY ITERATION
      NEND=30
      DO 5 N=1,NEND
      NUMIT=0
      TIME=TIME+DT
10    AMAX=0.
      NUMIT=NUMIT+1
      DO 15 I=2,22
      DO 15 J=2,22
      OLDVAL=HNEW(I,J)
      H1=(HOLD(I,J+1)+HOLD(I,J-1)+HOLD(I+1,J)+HOLD(I-1,J))/4.
      H2=(HNEW(I,J+1)+HNEW(I,J-1)+HNEW(I+1,J)+HNEW(I-1,J))/4.
      F1=DX*DX*S/(4.*T*DT)
      F2=1./(F1+ALPHA)
      HNEW(I,J)=((F1*HOLD(I,J))+(1.-ALPHA)*(H1-HOLD(I,J))+(ALPHA*H2)
     1+(R(I,J)*DX*DX/(4.*T)))*F2
      ERR=ABS(HNEW(I,J)-OLDVAL)
      IF(ERR.GT.AMAX) AMAX=ERR
15    CONTINUE
C     ADJUST NO FLOW BOUNDARIES
      DO 16 I=2,22
      HNEW(I,1)=HNEW(I,3)
      HNEW(I,23)=HNEW(I,21)
16    CONTINUE
      DO 17 J=2,22
      HNEW(1,J)=HNEW(3,J)
      HNEW(23,J)=HNEW(21,J)
```

(Continued)

[83]

Figure 4.6 (*Continued*)

```
54.       17 CONTINUE
55.          IF(ALPHA.LT.0.1) GO TO 18
56.          IF(AMAX.GT.TOL) GO TO 10
57.       18 CONTINUE
58. C     PREPARE FOR NEXT TIME STEP
59. C     PUT HNEW VALUES INTO HOLD ARRAY
60.          DO 20 I=1,23
61.          DO 20 J=1,23
62.       20 HOLD(I,J)=HNEW(I,J)
63. C     COMPUTE DRAWDOWN
64.          DO 25 I=2,22
65.          DO 25 J=2,22
66.          DD(I,J)=HO-HNEW(I,J)
67.       25 CONTINUE
68. C     PRINT RESULTS
69.          WRITE(6,29)  TIME,DT,NUMIT
70.       29 FORMAT(1X,'TIME=',F6.2,5X,'DT=',F6.2,5X,'NUMIT=',I4,/)
71.          WRITE(6,31)  ((DD(I,J),I=2,22),J=2,22)
72.       31 FORMAT (1X,21F5.2/)
73. C     INCREASE TIME STEP
74.          DT=DT*1.5
75.          IF(DT.GT.5.0)  DT=5.0
76.        5 CONTINUE
77.          STOP
78.          END
```

Figure 4.7
Sample output from the computer program in Figure 4.6 for time equals 13.1 days. Drawdowns are given in meters.

TIME= 13.12   DT= 4.38   NUMIT= 101

.50	.51	.52	.53	.56	.59	.63	.67	.73	.80	.88	.97	1.07	1.20	1.35	1.53	1.76	2.06	2.51	3.26	4.92	
.50	.51	.52	.53	.55	.58	.62	.67	.72	.78	.86	.94	1.04	1.16	1.29	1.45	1.64	1.86	2.10	2.35	2.80	3.26
.49	.50	.51	.52	.54	.57	.61	.65	.71	.77	.84	.92	1.01	1.12	1.24	1.37	1.52	1.69	1.86	2.00	2.10	2.07
.49	.50	.51	.52	.55	.58	.60	.64	.69	.75	.81	.88	.97	1.06	1.17	1.28	1.40	1.53	1.64	1.73	1.76	
.48	.49	.50	.52	.55	.58	.62	.67	.72	.78	.85	.92	1.00	1.09	1.18	1.28	1.37	1.45	1.51	1.53		
.46	.48	.49	.51	.53	.56	.60	.64	.69	.74	.80	.87	.94	1.01	1.09	1.17	1.24	1.30	1.34	1.35		
.45	.46	.48	.49	.52	.55	.58	.62	.66	.71	.76	.82	.88	.94	1.00	1.06	1.12	1.16	1.19	1.20		
.44	.44	.46	.48	.50	.52	.55	.59	.63	.67	.72	.77	.82	.87	.92	.97	1.01	1.05	1.07	1.08		
.43	.43	.45	.46	.48	.50	.53	.56	.60	.63	.67	.72	.76	.81	.85	.89	.92	.95	.96	.97		
.41	.41	.43	.44	.46	.48	.51	.53	.57	.60	.63	.67	.71	.75	.78	.81	.84	.86	.87	.88		
.40	.40	.41	.43	.44	.46	.48	.51	.54	.57	.60	.63	.66	.69	.72	.75	.77	.79	.80	.80		
.38	.39	.39	.40	.41	.43	.44	.46	.48	.51	.54	.56	.59	.62	.64	.67	.69	.71	.72	.73	.74	
.37	.37	.38	.39	.40	.41	.43	.45	.46	.48	.51	.53	.55	.57	.58	.60	.62	.64	.66	.67	.68	.68
.36	.36	.37	.37	.38	.40	.41	.43	.44	.46	.48	.50	.52	.53	.55	.57	.59	.60	.61	.62	.63	.63
.35	.35	.36	.36	.37	.38	.40	.41	.43	.44	.46	.48	.50	.51	.53	.54	.55	.56	.57	.58	.59	.59
.34	.34	.35	.35	.36	.37	.39	.40	.42	.43	.45	.46	.48	.49	.51	.52	.53	.54	.55	.56	.56	.56
.34	.34	.34	.35	.35	.36	.38	.39	.41	.42	.44	.45	.46	.48	.49	.50	.51	.52	.53	.53	.54	.54
.33	.33	.34	.34	.34	.35	.36	.38	.39	.40	.42	.43	.44	.45	.46	.47	.48	.49	.50	.50	.51	.51
.33	.33	.33	.34	.34	.35	.36	.37	.38	.39	.40	.42	.43	.44	.45	.46	.47	.48	.49	.50	.51	.51
.33	.33	.33	.33	.34	.34	.35	.36	.37	.38	.39	.40	.41	.42	.43	.44	.45	.46	.47	.48	.50	.51

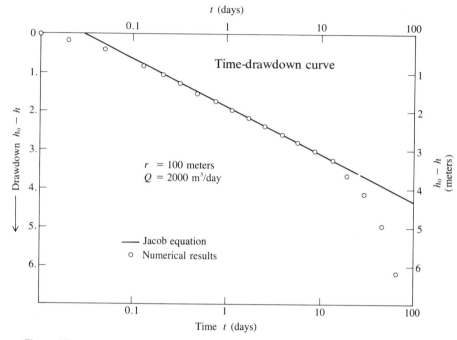

Figure 4.8
Comparison of the analytical solution using the Jacob equation with the numerical solution of the well drawdown example for $r = 100$ m. Deviation of the numerical solution from the Jacob time-drawdown plot after 13 days is due to boundary effects.

times is caused by boundary effects. After about 13 days, drawdowns at the boundaries are significant, and the numerical results deviate from the Theis solution. This situation is illustrated graphically in Figure 4.8, where the numerical solution using the Crank-Nicolson method is compared with the Jacob approximation of the Theis solution.

$$h_0 - h = \frac{Q}{4\pi T} \ln\left(\frac{2.25Tt}{r^2 S}\right) \tag{4.21}$$

In the Jacob approximation, the well function is approximated by $W(u) = -0.5772 - \ln u$, where u is less than about 0.01. Early numerical results deviate from the Jacob approximation for small values of t. After 13 days, the numerical results deviate from the Jacob approximation because of boundary effects.

4.4 UNCONFINED AQUIFER WITH DUPUIT ASSUMPTIONS

To derive the governing equation for transient unconfined flow, we must incorporate into Equation 3.14 a term to account for changes in storage. As in Section 4.1, this term is $S(\partial h/\partial t)$, where S is the unconfined storage coefficient defined by Equation 4.1. The governing equation for transient flow in an unconfined aquifer is

$$\frac{K}{2}\left(\frac{\partial^2 h^2}{\partial x^2} + \frac{\partial^2 h^2}{\partial y^2}\right) = S\frac{\partial h}{\partial t} - R(x, y, t) \tag{4.22}$$

The storage coefficient for an unconfined aquifer is also called specific yield. It is 100 to 1000 times greater than for a confined aquifer because changes in storage result from dewatering or saturating an aquifer, whereas, in a confined aquifer, water is released because of compression of the aquifer and, to a lesser degree, because of expansion of water in response to a decrease in pressure.

We again make the change of variable $v = h^2$. Because $\partial v/\partial t = \partial h^2/\partial t = 2h(\partial h/\partial t)$, we can rewrite Equation 4.22 as

$$\frac{K}{2}\left(\frac{\partial^2 v}{\partial x^2} + \frac{\partial^2 v}{\partial y^2}\right) = \frac{S}{2\sqrt{v}}\frac{\partial v}{\partial t} - R(x, y, t) \tag{4.23}$$

Explicit Approximation

We can now write the finite difference approximation in terms of v. First, we consider the explicit approximation with $\Delta x = \Delta y = a$. Equation 4.23 can be cast into the same form as Equation 4.3. By a derivation similar to that for Equation 4.8, the explicit approximation for v_{ij}^{n+1} is

$$v_{ij}^{n+1} = (1 - \omega)v_{ij}^n + \omega\left(\frac{v_{i+1,j}^n + v_{i-1,j}^n + v_{i,j+1}^n + v_{i,j-1}^n}{4}\right) + \frac{2R_{ij}\sqrt{v_{ij}^n}\,\Delta t}{S} \tag{4.24}$$

where

$$\omega = \frac{4K\sqrt{v_{ij}^n}\,\Delta t}{Sa^2}$$

Equation 4.24 has been linearized because a known value $\sqrt{v_{ij}^n}$ has been substituted for the nonlinear term. As with all explicit approximations, the ratio $K\sqrt{v_{ij}^n}\,\Delta t/Sa^2$ must be kept sufficiently small for the solution scheme to be stable.

Implicit Approximation

If we use the notation introduced by Equation 4.14 in Section 4.3, we can write an implicit approximation for Equation 4.23.

$$\alpha(\tilde{v}_{ij}^{n+1} - v_{ij}^{n+1}) + (1 - \alpha)(\tilde{v}_{ij}^{n} - v_{ij}^{n}) = \frac{a^2 S}{4K\sqrt{v_{ij}^{n}}} \frac{v_{ij}^{n+1} - v_{ij}^{n}}{\Delta t} - \frac{a^2 R_{ij}}{2K} \quad (4.25)$$

Solving Equation 4.25 for v_{ij}^{n+1} yields a Gauss–Seidel iteration formula analogous to Equation 4.17.

$$v_{ij}^{n+1} = \frac{1}{\left(\dfrac{a^2 S}{4K\sqrt{v_{ij}^{n}}\,\Delta t} + \alpha\right)} \left[\alpha\tilde{v}_{ij}^{n+1} + \frac{a^2 S}{4K\sqrt{v_{ij}^{n}}\,\Delta t} v_{ij}^{n} + (1 - \alpha)(\tilde{v}_{ij}^{n} - v_{ij}^{n}) + \frac{a^2 R_{ij}^{n}}{2K} \right]$$

$$(4.26)$$

The program in Figure 4.6 can be modified to use Equation 4.26 to solve transient flow problems in unconfined aquifers.

Additional Reading

1. For a more general derivation of the transient groundwater flow equation, see Remson et al. (1971, pp. 37–46) and Freeze and Cherry (1979, pp. 531–533).

2. For additional discussion of explicit and implicit approximations, see Remson et al. (1971, pp. 70–71 and 77–86), Bennett (1976, pp. 136–140), and Rushton and Redshaw (1979, pp. 165–171).

3. See Remson et al. (1971, pp. 71–77) for a more rigorous treatment of convergence and stability.

Problems

4.1 Compare the numerical solution of the reservoir example given in Figure 4.3 with the analytical solution from Carslaw and Jaeger (1959, p. 100):

$$h(x, t) = h_1 + \frac{(h_2 - h_1)x}{\ell} + \frac{2}{\pi} \sum_{n=1}^{\infty} \left[\frac{(h_2 - h_1) \cos(n\pi)}{n} \sin \frac{n\pi x}{\ell} e^{-Tn^2\pi^2 t/S\ell^2} \right] \quad (4.27)$$

where $h_1 = 16$ m, $h_2 = 11$ m, $\ell = 100$ m, $T = 0.02$ m^2 min^{-1}, and $S = 0.002$.

4.2 Given the drawdowns in Figure 4.7, construct a distance drawdown graph using the Jacob formula (Equation 4.21). Does the Crank–Nicolson solution agree with the Jacob approximation for $r > 100$ m at $t = 13.1$ days?

4.3 Suppose there is a drought on the island considered in Problem 3.4, and there is no recharge to the aquifer for a period of 300 days (that is, $R = 0$). Construct a graph showing the decline in water level in the well at the center of the island. Modify your program for solving Problem 3.4 to handle the transient flow equation with Dupuit assumptions using the explicit finite difference formula (Equation 4.24). (*Hint*: First modify your program to incorporate successive over relaxation (see Equation 2.12), and then set $\omega = 4K \Delta t \sqrt{v_{ij}^n}/Sa^2$.) Use the steady-state values of $h(x, y)$ that you generated in Problem 3.4 as initial head values and let $S = 0.2$. Add a mass balance calculation to your program and test the effect of changing Δt.

4.4 (a) Suppose the aquifer considered in the example in Section 4.2 is unconfined. Formulate a mathematical model (that is, state the governing equation, boundary conditions, and initial conditions) to solve for the water table profile through time if, at $t = 0$, the water level of 16 m in one of the reservoirs is suddenly dropped to 11 m (see Figure 4.1).

Modify the computer program in Figure 4.2 to solve the one-dimensional form of the explicit approximation for flow through an unconfined aquifer. Your program should use the one-dimensional form of Equation 4.24. Set $R = 0$ and let $S = 0.2$ and $K = 0.002$ m min^{-1}. Construct a graph similar to the one in Figure 4.4a. Estimate the maximum permissible value of Δt necessary to ensure stability by setting $4K \sqrt{v_{ij}^n} \Delta t/Sa^2 = 1$ and solving for Δt. Check the numerical steady-state solution with the analytical solution.

$$h^2 = \frac{h_2^2 - h_1^2}{\ell} x + h_1^2 \quad (4.28)$$

(b) Suppose the aquifer is being recharged from precipitation at a rate of 1.2×10^{-5} m min^{-1}. Solve for the water table profile through time if, at $t = 0$, the water level of 16 m in one of the reservoirs is suddenly dropped to 11 m. For initial conditions, use the steady-state water table profile where $R = 1.2 \times 10^{-5}$ m min^{-1} and $h_1 = h_2 = 16$ m. Construct a graph of the water table profile through time, similar to Figure 4.4a, and check the numerical steady-state solution with the analytical solution.

$$h^2 = -\frac{R}{K} x^2 + \left(\frac{h_2^2 - h_1^2}{\ell} + \frac{R\ell}{K} \right) x + h_1^2 \quad (4.29)$$

4.5 Consider the plan view of the unconfined aquifer shown in Figure 4.9. A mesh-centered grid system has been overlain such that $\Delta x = \Delta y = 100$ m. The boundary conditions are also given. The heads are relative to a datum at the base of the aquifer. The shaded area in the figure represents a gravel pit which is to be dewatered by pumping from 12 wells placed along the boundary of the pit. Each well is to be pumped at a rate of 3500 m³ day⁻¹. The hydraulic conductivity of the aquifer is 50 m day⁻¹ and $S = 0.2$.

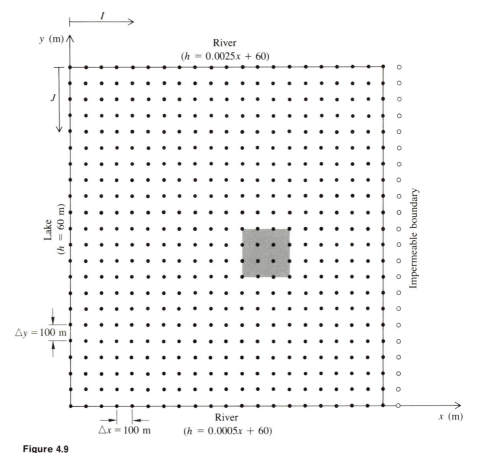

Figure 4.9
Finite difference grid for the dewatering problem. Open circles represent fictitious nodes used to simulate the no-flow boundary. Shaded area represents the gravel pit surrounded by 12 wells, which are located at each of the 12 nodes along the boundary of the shaded area.

(a) Modify the computer program in Figure 4.6 to compute the drawdowns in the gravel pit after 200 days of pumping. For initial conditions, use the steady-state head configuration without pumping.

(b) Write a mass balance calculation routine and incorporate it into your program. You will need to calculate the volume of water removed by pumping at the 12 nodes, the volume of water removed from storage over the entire aquifer, and the volume of water flowing across the boundaries. Until the cone of depression intersects a boundary, inflow across the boundaries should equal outflow across the boundaries because all water being pumped is removed from storage within the aquifer. When the cone of depression spreads to the recharge boundaries, the volume of water derived from storage will be less than the amount pumped, indicating that the rest of the water being pumped from the aquifer is derived from the recharge boundaries.

(c) Modify the program in Figure 3.12 to generate the steady-state solution after dewatering. Compare this steady-state solution with the solution after 200 days of pumping, which you computed in part (a).

CHAPTER 5

Other Solution Methods

5.1 INTRODUCTION

The computer programs presented in this book are relatively simple because they are designed to solve relatively simple problems. For most field applications, a model with more flexibility is desired. For example, it is often necessary to use a model which allows for consideration of anisotropic and heterogeneous aquifers. Also, for many problems it is convenient to allow for irregular finite difference grids where Δx and Δy are not constants. While iterative techniques are adequate for solving relatively simple problems, more complex models are solved more efficiently through the use of direct methods in combination with iteration.

When iterative methods are used, a single generalized finite difference expression is written and then solved repetitively for every node in the problem domain. When direct methods are used to solve a set of linear algebraic equations, a sequence of operations is performed only once. In this chapter, we introduce direct methods and discuss certain numerical techniques utilizing direct methods in combination with iteration. These techniques are incorporated in two of the most commonly used and well-documented generic finite difference models—the Prickett–Lonnquist model (1971) and the Trescott–Pinder–Larson model (1976). We also describe the procedure for calibrating and verifying models used to study field problems.

5.2   MATRIX NOTATION

Matrix notation simplifies the task of communicating the manipulations to be performed on a set of linear equations. The finite element technique described in Chapters 6 and 7 also leads to a set of linear equations, and the development of this technique is greatly facilitated by the use of matrix notation. Some familiarity with the rudiments of matrix notation and multiplication is assumed.

Consider the following set of linear equations.

$$
\begin{aligned}
x_1 + 2x_2 - 3x_3 &= 5 \\
-2x_1 + 7x_2 + x_3 &= 11 \\
3x_2 - x_3 &= 3
\end{aligned}
$$

This set of equations is written in matrix form as

$$
\begin{bmatrix} 1 & 2 & -3 \\ -2 & 7 & 1 \\ 0 & 3 & -1 \end{bmatrix} \begin{Bmatrix} x_1 \\ x_2 \\ x_3 \end{Bmatrix} = \begin{Bmatrix} 5 \\ 11 \\ 3 \end{Bmatrix}
$$

The square matrix consists of the coefficients of the unknowns x_1, x_2, x_3. This coefficient matrix is represented by square brackets. The unknowns and knowns are column matrices. They are represented by braces.

A set of n linear equations in n unknowns can be written in the general form

$$
\begin{aligned}
A_{1,1}x_1 + A_{1,2}x_2 + \cdots + A_{1,n}x_n &= f_1 \\
A_{2,1}x_1 + A_{2,2}x_2 + \cdots + A_{2,n}x_n &= f_2 \\
&\vdots \\
A_{n,1}x_1 + A_{n,2}x_2 + \cdots + A_{n,n}x_n &= f_n
\end{aligned}
\tag{5.1}
$$

This set of linear equations is written in matrix form as

$$
\begin{bmatrix} A_{1,1} & A_{1,2} & \cdots & A_{1,n} \\ A_{2,1} & A_{2,2} & \cdots & A_{2,n} \\ & & \vdots & \\ A_{n,1} & A_{n,2} & \cdots & A_{n,n} \end{bmatrix} \begin{Bmatrix} x_1 \\ x_2 \\ \vdots \\ x_n \end{Bmatrix} = \begin{Bmatrix} f_1 \\ f_2 \\ \vdots \\ f_n \end{Bmatrix}
\tag{5.2}
$$

The matrix equation can be rewritten in a more simplified notation as

$$[A]\{x\} = \{f\} \tag{5.3}$$

where $[A]$, $\{x\}$, and $\{f\}$ stand for the complete matrices written out fully in Equation 5.2. A typical entry $A_{i,j}$ in the matrix $[A]$ is the coefficient in the ith row and jth column. Notice that the index convention in matrix notation is different from the one we have used for identifying nodes in previous chapters.

The set of algebraic equations, Equation 5.1, is generated by matrix multiplication of $[A]$ and $\{x\}$. For example, to generate the first algebraic equation, each entry $A_{1,j}$ in the first row of the coefficient matrix is multiplied by a column entry x_j in the $\{x\}$ array, and the products are summed and set equal to the first entry f_1 in the $\{f\}$ array. In general, the products $A_{i,j}x_j$ are summed and set equal to f_i.

5.3 TRIDIAGONAL MATRICES

Consider the one-dimensional transient flow equation.

$$\frac{\partial^2 h}{\partial x^2} = \frac{S}{T}\frac{\partial h}{\partial t} \tag{5.4}$$

The implicit or backward finite difference approximation, where the space derivative is evaluated at the advanced time level $(n + 1)$, is

$$\frac{h_{i-1}^{n+1} - 2h_i^{n+1} + h_{i+1}^{n+1}}{(\Delta x)^2} = \frac{S}{T}\frac{h_i^{n+1} - h_i^n}{\Delta t} \tag{5.5}$$

Suppose that we have a problem domain with six nodes where the first and last nodes are boundary nodes of known head. We wish to write the set of algebraic equations that would be generated by applying Equation 5.5 to these nodes, and we wish to write it in matrix form. First, we rearrange Equation 5.5 and put unknowns, that is, heads at the $(n + 1)$ time level, on the left-hand side and put knowns on the right-hand side.

$$h_{i-1}^{n+1} + \left(-2 - \frac{S(\Delta x)^2}{T\,\Delta t}\right)h_i^{n+1} + h_{i+1}^{n+1} = -\frac{S(\Delta x)^2}{T\,\Delta t}h_i^n \tag{5.6}$$

If the head values h_1 and h_6, which are known from the boundary conditions, are transferred to the right-hand side, then the matrix form of the set of algebraic equations for the six-node problem is

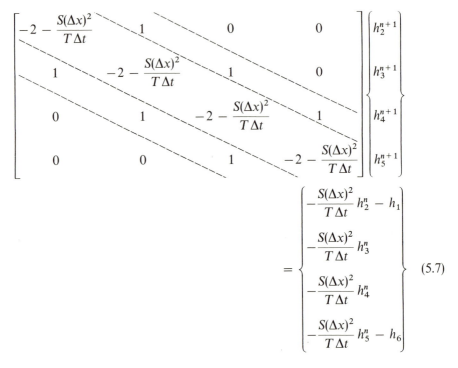

$$(5.7)$$

The coefficient matrix has nonzero entries only along the three center diagonals. This type of matrix is known as a tridiagonal matrix.

Thomas Algorithm

In general, a matrix equation such as Equation 5.2 can be solved by a technique known as Gaussian elimination. However, except for special types of coefficient matrices, Gaussian elimination requires a great deal of computer storage and computer time. A particularly efficient form of Gaussian elimination can be used to solve matrix equations which have a tridiagonal coefficient matrix. This form of Gaussian elimination utilizes the Thomas algorithm. Remson et al. (1971) present the details of both Gaussian elimination and the Thomas algorithm. We develop the Thomas algorithm for tridiagonal matrices by systematically solving the linear equations by row operations.

We begin with a set of four equations whose coefficients are in tridiagonal form.

$$b_1 x_1 + c_1 x_2 = f_1$$
$$a_2 x_1 + b_2 x_2 + c_2 x_3 = f_2$$
$$a_3 x_2 + b_3 x_3 + c_3 x_4 = f_3 \qquad (5.8)$$
$$a_4 x_3 + b_4 x_4 = f_4$$

The notation is that a_i is a subdiagonal coefficient, b_i is a center diagonal coefficient, and c_i is a superdiagonal coefficient. The subscript i indicates row number. We will perform row operations in a systematic manner to eliminate the subdiagonal terms and to normalize the coefficients of diagonal terms to 1. The idea is to transform the original tridiagonal set of equations into an equivalent upper diagonal set.

$$x_1 + \beta_1 x_2 = y_1$$
$$x_2 + \beta_2 x_3 = y_2$$
$$x_3 + \beta_3 x_4 = y_3 \qquad (5.9)$$
$$x_4 = y_4$$

To put Equation 5.8 into the form of Equation 5.9, we need to find expressions for β_i and y_i. We do row operations such as one would use in solving the system of equations by hand. If we compare the first row of Equation 5.9 with that of Equation 5.8, we see that $\beta_1 = c_1/b_1$ and $y_1 = f_1/b_1$. We can obtain the form of the second row of Equation 5.9 if we eliminate x_1 between the first row of Equation 5.9 and the second row of Equation 5.8. The first row of Equation 5.9 is multiplied by a_2 and subtracted from the second row of Equation 5.8. This produces the equation

$$(b_2 - a_2\beta_1)x_2 + c_2 x_3 = f_2 - a_2 y_1 \qquad (5.10)$$

We define the coefficient of x_2 in Equation 5.10 to be $\alpha_2 = b_2 - a_2\beta_1$. Dividing Equation 5.10 by α_2 gives

$$x_2 + \frac{c_2}{\alpha_2} x_3 = \frac{f_2 - a_2 y_1}{\alpha_2} \qquad (5.11)$$

Equation 5.11 is now in the form of the second row of Equation 5.9. We can therefore make the identification that $\beta_2 = c_2/\alpha_2$ and $y_2 = (f_2 - a_2 y_1)/\alpha_2$. By continuing down the rows in this fashion, we find the general relations

$$\alpha_i = b_i - a_i\beta_{i-1} \qquad (5.12)$$
$$\beta_i = c_i/\alpha_i \qquad (5.13)$$

and

$$y_i = (f_i - a_i y_{i-1})/\alpha_i \tag{5.14}$$

If β_0 and y_0 are defined to be equal to zero, the recursive expressions also hold for $i = 1$. All the coefficients β_i and knowns y_i in Equation 5.9 are now defined. The solution of the original problem, Equation 5.8, is done systematically by backward substitution from the bottom row $x_4 = y_4$ to the top row. In general, $x_n = y_n$, where n is the number of equations, and, for $i < n$,

$$x_i = y_i - \beta_i x_{i+1} \tag{5.15}$$

The recursive relations, Equations 5.12 to 5.15, constitute the Thomas algorithm. The Thomas algorithm is programmed in *SUBROUTINE TRIDIA* (Figure 5.1).

Figure 5.1
Subroutine *TRIDIA*. Subroutine for solving a matrix equation with a tridiagonal coefficient matrix using the Thomas algorithm.

```
 1.            SUBROUTINE TRIDIA(N)
 2.        C   THIS SUBROUTINE CONTAINS THE THOMAS ALGORITHM
 3.        C   THE SOLUTION IS CONTAINED IN THE X ARRAY
 4.            COMMON A,B,C,X,F
 5.            DIMENSION A(50),B(50),C(50),X(50),F(50)
 6.            DIMENSION ALPHA(50),BETA(50),Y(50)
 7.            ALPHA(1)=B(1)
 8.            BETA(1)=C(1)/ALPHA(1)
 9.            Y(1)=F(1)/ALPHA(1)
10.            DO 201 I=2,N
11.            ALPHA(I)=B(I)-A(I)*BETA(I-1)
12.            BETA(I)=C(I)/ALPHA(I)
13.     201    Y(I)=(F(I)-A(I)*Y(I-1))/ALPHA(I)
14.        C   BEGIN BACKWARD SUBSTITUTION FROM LAST ROW
15.            X(N)=Y(N)
16.            NU=N-1
17.            DO 203 I=1,NU
18.            J=N-I
19.     203    X(J)=Y(J)-BETA(J)*X(J+1)
20.            RETURN
21.            END
```

Direct Solution of Reservoir Problem

In Section 4.2, we introduced the one-dimensional transient example involving flow between two reservoirs. The problem was to determine the change in the potentiometric surface through time if the level of one of the reservoirs drops suddenly from 16 m above datum to 11 m above datum. In Chapter 4, we solved the problem using an explicit approximation. In order to ensure that the numerical solution remained stable, it was necessary to keep Δt less than or equal to 5 minutes. In this section, we consider the solution of this problem using an implicit approximation and a matrix solution technique based on the Thomas algorithm.

At the beginning of this section, we established that the fully implicit approximation to the one-dimensional transient flow equation yields a matrix equation whose coefficient matrix is tridiagonal. Such a matrix equation must be solved at each time step. We saw how the Thomas algorithm provides an efficient way of solving a matrix equation whose coefficient matrix is tridiagonal.

The computer program in Figure 5.2 utilizes the Thomas algorithm, as programmed in *SUBROUTINE TRIDIA*, to provide a direct solution to the reservoir example. The system of equations to be solved looks similar to Equation 5.7. In the main program, the entries in the $[A]$, $\{x\}$, and $\{f\}$ arrays are defined by the correspondence between the general tridiagonal form, Equation 5.8, and Equation 5.7. The number of unknowns is two fewer than the number of nodes because the two boundary nodes have specified heads. The first unknown is h_2, and it corresponds to x_1 in *SUBROUTINE TRIDIA*. In general, $h_i = x_{i-1}$, and this correspondence is made in line 43 of Figure 5.2, right after the call to *TRIDIA*.

If we compare this program with the explicit approximation program (Figure 4.2), we observe that we do not need two head arrays, *HNEW* and *HOLD*, because old or known head values are incorporated into the entries of the *F* array. Also, because the fully implicit approximation is used, it is unnecessary to limit the size of Δt. In Figure 5.2, Δt is set equal to 10 minutes and heads are printed after every time step.

5.4 ALTERNATING DIRECTION IMPLICIT (ADI) METHOD

We have seen that if the coefficient matrix is tridiagonal, the Thomas algorithm provides a way of solving the matrix equation in an efficient manner. Although the fully implicit approximation led to a tridiagonal coefficient matrix for the one-dimensional transient flow equation, it does not yield a tridiagonal coefficient matrix for the two-dimensional transient flow equation.

Figure 5.2
Computer program for solving the reservoir example of Section 4.2 using subroutine
TRIDIA (line 40).

```
1.    C   DIRECT SOLUTION (THOMAS ALGORITHM) OF 1-D TRANSIENT FLOW EQN.
2.    C   RESERVOIR PROBLEM
3.    C   ARRAYS A,B,C CONTAIN THE ENTRIES IN TRIDIAGONAL COEFFICIENT MATRIX
4.    C   NX IS THE TOTAL NUMBER OF NODES INCLUDING BOUNDARY NODES
5.    C   NEND IS THE NUMBER OF TIME STEPS
6.            COMMON A,B,C,X,F
7.            DIMENSION A(5Ø),B(5Ø),C(5Ø),X(5Ø),F(5Ø),H(5Ø)
8.            DT=1Ø.
9.            DELX=1Ø.
1Ø.           T=Ø.Ø2
11.           S=Ø.ØØ2
12.           NX=11
13.           NLX=NX-1
14.           N=NX-2
15.   C   INITIAL AND BOUNDARY CONDITIONS FOR HEAD ARRAY, H
16.           HO=16.
17.           DO 2Ø I=1,NLX
18.           H(I)=HO
19.   2Ø      CONTINUE
2Ø.           H(NX)=11.
```

```
21.        C    DEFINE ENTRIES IN THE COEFFICIENT MATRIX
22.              FAC1=-S*DELX*DELX/(T*DT)
23.              FAC2=-2.+FAC1
24.              DO 10 I=1,N
25.              A(I)=1.
26.              C(I)=1.
27.              B(I)=FAC2
28.        10    CONTINUE
29.        C    COMPUTE HEADS THROUGH TIME
30.              TIME=DT
31.              NEND=50
32.              DO 50 NE=1,NEND
33.        C    DEFINE THE ENTRIES IN THE F ARRAY FOR EACH TIME STEP
34.              DO 30 I=1,N
35.              F(I)=FAC1*H(I+1)
36.        30    CONTINUE
37.              F(1)=F(1)-H(1)
38.              F(N)=F(N)-H(NX)
39.        C    SOLVE THE SET OF EQUATIONS USING THE THOMAS ALGORITHM
40.              CALL TRIDIA(N)
41.        C    UNKNOWNS X(1) TO X(N) CORRESPOND TO H(2) TO H(NLX)
42.              DO 45 I=2,NLX
43.        45    H(I)=X(I-1)
44.              PRINT 40,(H(I),I=1,NX),TIME
45.        40    FORMAT(1X,11F6.2,F10.2)
46.              TIME=TIME+DT
47.        50    CONTINUE
48.              STOP
49.              END
```

[101]

Consider the two-dimensional transient equation for a confined aquifer.

$$\frac{\partial^2 h}{\partial x^2} + \frac{\partial^2 h}{\partial y^2} = \frac{S}{T} \frac{\partial h}{\partial t} \tag{5.16}$$

For $\Delta x = \Delta y = a$, the fully implicit finite difference approximation is

$$h^{n+1}_{i+1,j} + h^{n+1}_{i-1,j} + h^{n+1}_{i,j+1} + h^{n+1}_{i,j-1} - 4h^{n+1}_{ij} = \frac{Sa^2}{T} \frac{h^{n+1}_{ij} - h^n_{ij}}{\Delta t} \tag{5.17}$$

Putting unknown heads on the left-hand side of the equation and known heads on the right-hand side yields

$$h^{n+1}_{i+1,j} + h^{n+1}_{i-1,j} + h^{n+1}_{i,j+1} + h^{n+1}_{i,j-1} + \left(-4 - \frac{Sa^2}{T\Delta t}\right) h^{n+1}_{ij} = -\frac{Sa^2}{T\Delta t} h^n_{ij} \tag{5.18}$$

If we applied Equation 5.18 to a four-by-four system of sixteen nodes where the twelve boundary heads are known (Figure 5.3), the resulting set of equations in matrix form is

$$\begin{bmatrix} -4 - \dfrac{Sa^2}{T\Delta t} & 1 & 1 & 0 \\[2ex] 1 & -4 - \dfrac{Sa^2}{T\Delta t} & 0 & 1 \\[2ex] 1 & 0 & -4 - \dfrac{Sa^2}{T\Delta t} & 1 \\[2ex] 0 & 1 & 1 & -4 - \dfrac{Sa^2}{T\Delta t} \end{bmatrix} \begin{Bmatrix} h^{n+1}_{2,2} \\[2ex] h^{n+1}_{3,2} \\[2ex] h^{n+1}_{2,3} \\[2ex] h^{n+1}_{3,3} \end{Bmatrix}$$

$$= \begin{Bmatrix} -\dfrac{Sa^2}{T\Delta t} h^n_{2,2} - h_{1,2} - h_{2,1} \\[2ex] -\dfrac{Sa^2}{T\Delta t} h^n_{3,2} - h_{3,1} - h_{4,2} \\[2ex] -\dfrac{Sa^2}{T\Delta t} h^n_{2,3} - h_{1,3} - h_{2,4} \\[2ex] -\dfrac{Sa^2}{T\Delta t} h^n_{3,3} - h_{4,3} - h_{3,4} \end{Bmatrix} \tag{5.19}$$

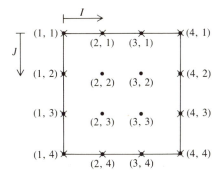

Figure 5.3
Sixteen node example used to illustrate the ADI method. The symbol ✕ represents a constant head boundary node.

Note that the coefficient matrix is not tridiagonal. It is a banded symmetric matrix and can be solved directly, but we cannot use the Thomas algorithm.

The basis of the Alternating Direction Implicit (ADI) method is to obtain a tridiagonal coefficient matrix by alternately writing the finite difference equations, first implicitly along columns and explicitly along rows, and then vice versa. In the first step of such a pair of steps, heads along columns are approximated at time level $(n + 1)$, and heads along rows are approximated at time level (n). Then, Equation 5.18 becomes

$$h_{i,j-1}^{n+1} + \left(-4 - \frac{Sa^2}{T\,\Delta t}\right)h_{ij}^{n+1} + h_{i,j+1}^{n+1} = -\frac{Sa^2}{T\,\Delta t}h_{ij}^{n} - h_{i+1,j}^{n} - h_{i-1,j}^{n} \quad (5.20)$$

Comparison of Equation 5.20 with Equation 5.6 shows that Equation 5.20 will yield a tridiagonal coefficient matrix along any column. For each column, we have a set of equations for the heads in that column and no other. That is, the explicit approximation along rows uncouples one column from another. Specifically, for the nodal array in Figure 5.3, we can formulate two one-dimensional problems oriented along the second and third columns (Figure 5.4a). We use Equation 5.20 to generate two sets of algebraic equations—one set for each column. In matrix form, the set of equations for the second column of nodes in Figure 5.4a is

$$\begin{bmatrix} -4 - \dfrac{Sa^2}{T\,\Delta t} & 1 \\ 1 & -4 - \dfrac{Sa^2}{T\,\Delta t} \end{bmatrix} \begin{Bmatrix} h_{2,2}^{n+1} \\ h_{2,3}^{n+1} \end{Bmatrix} = \begin{Bmatrix} -\dfrac{Sa^2}{T\,\Delta t}h_{2,2}^{n} - h_{3,2}^{n} - h_{1,2} - h_{2,1} \\ -\dfrac{Sa^2}{T\,\Delta t}h_{2,3}^{n} - h_{3,3}^{n} - h_{2,4} - h_{1,3} \end{Bmatrix} \quad (5.21)$$

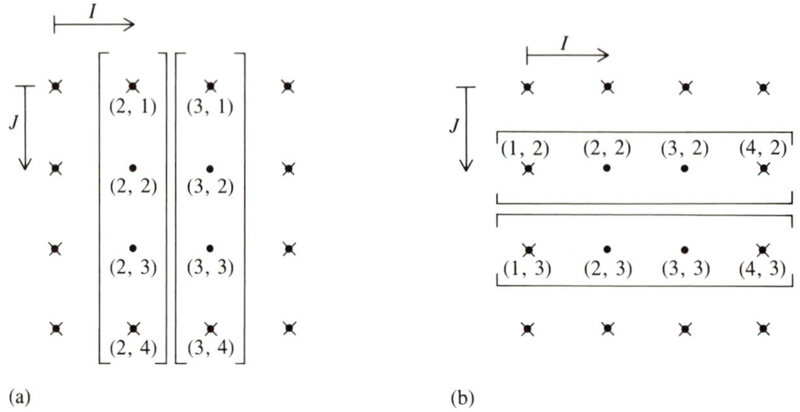

Figure 5.4
Sixteen node example illustrating the solution procedure used in the ADI method.
(a) Solution oriented along columns. (b) Solution oriented along rows.

In matrix form, the set of equations for the third column of nodes in Figure 5.4a is

$$
\begin{bmatrix}
-4 - \dfrac{Sa^2}{T\,\Delta t} & 1 \\[3mm]
1 & -4 - \dfrac{Sa^2}{T\,\Delta t}
\end{bmatrix}
\begin{Bmatrix}
h^{n+1}_{3,2} \\[3mm]
h^{n+1}_{3,3}
\end{Bmatrix}
=
\begin{Bmatrix}
-\dfrac{Sa^2}{T\,\Delta t}\, h^n_{3,2} - h^n_{2,2} - h_{3,1} - h_{4,2} \\[3mm]
-\dfrac{Sa^2}{T\,\Delta t}\, h^n_{3,3} - h^n_{2,3} - h_{4,3} - h_{3,4}
\end{Bmatrix}
\tag{5.22}
$$

Each matrix equation can be solved separately using the Thomas algorithm. For a problem with thirty five columns, two of which contain boundary nodes of known head, we would generate thirty three matrix equations and use the Thomas algorithm thirty three times.

This first step of ADI has uncoupled the problem by columns. The separate matrix equation for each column of nodes represents a partitioning of the system of equations involving all the unknowns. Thus the left-hand side of the full system of equations equivalent to Equations 5.21 and 5.22 is

$$
\begin{bmatrix}
-4 - \dfrac{Sa^2}{T\,\Delta t} & 1 & \bigm| & 0 & 0 \\[2ex]
1 & -4 - \dfrac{Sa^2}{T\,\Delta t} & \bigm| & 0 & 0 \\[1ex]
\hline
0 & 0 & \bigm| & -4 - \dfrac{Sa^2}{T\,\Delta t} & 1 \\[2ex]
0 & 0 & \bigm| & 1 & -4 - \dfrac{Sa^2}{T\,\Delta t}
\end{bmatrix}
\begin{Bmatrix}
h_{2,2}^{n+1} \\[1ex]
h_{2,3}^{n+1} \\[1ex]
h_{3,2}^{n+1} \\[1ex]
h_{3,3}^{n+1}
\end{Bmatrix}
$$

The four-by-four coefficient matrix is also tridiagonal, and the whole system could be solved by a single use of the Thomas algorithm. However, the partitioning into several smaller tridiagonal problems reduces the sizes of the auxiliary arrays required in *SUBROUTINE TRIDIA*.

We return to our example in Figure 5.3 for computation of heads at the next time step using ADI. After we have solved our two matrix equations, Equations 5.21 and 5.22, we have new values for each unknown head. For the next time step, the one-dimensional problems are oriented along rows (Figure 5.4b). We retain the notation that (n) represents the time level of known heads and $(n + 1)$ represents the next time level. Then, the finite difference equation analogous to Equation 5.20 for the case where the heads are written implicitly along rows and explicitly along columns is

$$
h_{i-1,j}^{n+1} + \left(-4 - \frac{Sa^2}{T\,\Delta t} \right) h_{ij}^{n+1} + h_{i+1,j}^{n+1} = -\frac{Sa^2}{T\,\Delta t} h_{ij}^{n} - h_{i,j+1}^{n} - h_{i,j-1}^{n} \quad (5.23)
$$

The explicit approximation along columns uncouples one row from another. For each row, we have a set of equations for heads in that row and no other. Therefore, Equation 5.23 will also generate a set of matrix equations—one for each interior row—with tridiagonal coefficient matrices.

Alternating the explicit approximation between columns and rows is an attempt to compensate for errors generated in either direction. Therefore, the length of the time step should not be changed within a pair of steps, and the total number of time steps should be an even number.

Iterative ADI

The ADI method as just described sometimes leads to a large error at particular nodes or time levels (Bennett, 1976, p. 149). These errors can be reduced by combining the ADI method with iteration. Prickett and Lonnquist (1971), for example, use a version of iterative ADI which incorporates Gauss–Seidel iteration. Within a single time step, the solution is oriented alternately along columns and rows until the heads converge within a specified error tolerance.

The sequence of finite difference equations for several Gauss–Seidel iterations within a time step, where (m) is the iteration index, is as follows. From iteration level (m) to $(m + 1)$, the solution is oriented along columns. The unknowns are the heads in column i at time step $(n + 1)$.

$$h_{i,j-1,n+1}^{m+1} + \left(-4 - \frac{Sa^2}{T\,\Delta t} \right) h_{i,j,n+1}^{m+1} + h_{i,j+1,n+1}^{m+1}$$

$$= -\frac{Sa^2}{T\,\Delta t} h_{i,j,n} - h_{i+1,j,n+1}^{m} - h_{i-1,j,n+1}^{m+1} \quad (5.24)$$

The time index n now occupies a subscript position, and the iteration index m takes the superscript position. In Equation 5.24, explicitly defined values of head are used for heads not in column i. These values in columns $(i - 1)$ and $(i + 1)$ are brought to the right-hand side of Equation 5.24. In Gauss–Seidel iteration, any updated values are immediately put to use within the same iteration level. Therefore, note that in Equation 5.24, $h_{i-1,j,n+1}$ is approximated at the current iteration level $(m + 1)$ because it was computed when the system of equations involving column $(i - 1)$ was solved. However, $h_{i+1,j,n+1}$ is approximated at the old iteration level (m) because it is in the next column $(i + 1)$, which is yet to be computed as the columns are processed one by one.

Next, from iteration level $(m + 1)$ to $(m + 2)$, the solution is oriented along rows. The unknowns are the heads in row j at time step $(n + 1)$.

$$h_{i-1,j,n+1}^{m+2} + \left(-4 - \frac{Sa^2}{T\,\Delta t} \right) h_{i,j,n+1}^{m+2} + h_{i+1,j,n+1}^{m+2}$$

$$= -\frac{Sa^2}{T\,\Delta t} h_{i,j,n} - h_{i,j+1,n+1}^{m+1} - h_{i,j-1,n+1}^{m+2} \quad (5.25)$$

The process recycles. From iteration level $(m + 2)$ to $(m + 3)$, the equations are written along columns, and, from iteration level $(m + 3)$ to $(m + 4)$, the equations are written along rows. At each iteration level, we must solve a set of matrix equations using the Thomas algorithm. The convergence is checked after every other iteration because a complete iteration cycle consists of two iteration levels—one oriented along columns and one oriented along rows.

Iterative ADI can be summarized as follows. In the first iteration level of a cycle, a set of linear equations is generated for each individual column of nodes. Column by column, the equations are sequentially generated and solved. In the second iteration level, a set of linear equations is generated for each individual row of nodes. Row by row, the equations are sequentially generated and solved. The two steps are repeated until convergence is achieved. Thus, to advance a single time step using iterative ADI requires the solution of many small sets of linear equations rather than a single large set.

5.5   PRICKETT–LONNQUIST AND TRESCOTT–PINDER–LARSON MODELS

We have dealt with relatively simple groundwater systems to illustrate the application of numerical methods. For most regional field problems, however, it is necessary to simulate anisotropic and heterogeneous aquifers. Furthermore, it is usually advantageous to use an irregular finite difference mesh where Δx and Δy are not constants. In general, one-dimensional models are not valid for field applications, but, for most problems, there will not be enough field data to warrant the use of a three-dimensional model. In fact, for most field problems, two-dimensional models are adequate. Two-dimensional models may be either areal models, where the vertical component of flow is assumed to be negligible, or profile models, where flow along one of the horizontal axes is assumed to be zero.

The two most widely used and well-documented, two-dimensional finite difference models are the groundwater flow models developed by Prickett and Lonnquist (1971) and Trescott et al. (1976). Both models solve a form of the transient flow equation which allows for the inclusion of heterogeneity and anisotropy.

$$\frac{\partial}{\partial x}\left(T_x \frac{\partial h}{\partial x}\right) + \frac{\partial}{\partial y}\left(T_y \frac{\partial h}{\partial y}\right) = S \frac{\partial h}{\partial t} - R(x, y, t) \qquad (5.26)$$

Although strictly valid only for confined aquifers, Equation 5.26 can also be used for unconfined aquifers by allowing T_x and T_y to vary with time as the saturated thickness changes. The numerical models can also be used to solve for steady-state conditions by specifying a large time step (Prickett and Lonnquist, 1971) or by setting the storage coefficient equal to zero (Trescott et al., 1976).

Prickett and Lonnquist (1971) use the ADI method in combination with Gauss–Seidel iteration as the solution procedure. Trescott et al. (1976) present three options for solving their numerical model, one of which is iterative ADI, but, instead of using Gauss–Seidel iteration, they use a set of iteration parameters. Details on the use and method of calculating the parameters for each iteration cycle are given by Trescott et al. (1976). The other two solution procedures used by Trescott et al. (1976) are discussed in the notes at the end of this chapter.

Both models allow for Δx and Δy to vary areally, and both use a block-centered grid. We use a mesh-centered grid in our examples. Faust and Mercer (1980a, p. 396) point out that "the choice of the type of grid to use depends largely on the boundary conditions. The mesh-centered grid is convenient for problems where the values of head are specified on the boundary, whereas the block-centered grid has an advantage in problems where the flux is specified across the boundary. From a practical point of view, the differences in the two types of grids are minor."

Input Data

The input necessary to solve a model of the kind developed by Prickett and Lonnquist (1971) and Trescott et al. (1976) consists of the following types of information:

1. Definition of boundary conditions.

2. Values of T_x, T_y, S, and R for the area of influence of each node.

3. Initial guesses of head for steady-state simulations or initial conditions (that is, values of head for each node) for transient simulations.

Definition of boundary conditions is one of the most difficult problems in designing a model. Natural boundaries, such as major groundwater recharge or discharge areas or impermeable boundaries, may be located too far from the modeled area to be included. A common practice is to make the modeled area large enough so that errors in the boundary conditions will have little effect on the heads in the interior portion of the grid. If this is done, the sensitivity of the model to changes in boundary conditions should be determined. Of course, physically realistic boundary conditions should be used whenever possible.

It is uncommon to have field determined values of T_x, T_y, S, and R for each node in the system. Usually, only a few values for T and S will be available from pumping test data for the entire modeled area. This information is used together with information on stratigraphy and specific capacity (the ratio of discharge rate to drawdown) from well logs to estimate the distribution of T_x, T_y, and S over the grid. Values of areal recharge rate can only be determined roughly based on the amount of precipitation and evapotranspiration. Pumping rates of wells must also be considered when specifying values for R.

For steady-state simulations, any set of initial guesses for head can be selected. However, it should be remembered that the better the guesses, the fewer the number of iterations that will be required for the solution to converge. For transient simulations, it is common to use a steady-state head configuration as the initial conditions. Trescott et al. (1976, p. 30) point out that "if initial conditions are specified so that transient flow is occurring in the system at the start of the simulation, it should be recognized that water levels will change during the simulation, not only in response to the new pumping stress, but also due to the initial conditions. This may or may not be the intent of the user."

5.6 CALIBRATION AND VERIFICATION

Calibration

In general, the first step in model calibration is to design a steady-state model to solve for the head distribution to be used as the initial conditions in a later transient simulation. Given the input data discussed in the previous section, it is possible to generate a solution for the heads at each nodal point. However, in order to verify the accuracy of the solution, it is necessary to match the computed heads with heads measured at a number of points in the field. Invariably, the heads computed from the first run of the model will not match the field values. Calibration consists of adjusting the input data until computed heads match the field values. It is not unreasonable to adjust the input data because these data are imperfectly known, and there will be a certain range of values that may be valid. It is not uncommon to make from twenty to fifty trial-and-error simulations before an acceptable calibration is achieved. Calibration is really a way of solving the inverse problem discussed in Section 3.4.

Novice modelers are often tempted to use the measured values of head as input to the model. That is, the temptation is to force heads in the interior of the grid to equal the field measured values in the hope that this will ensure calibration. However, if the boundary conditions and values for T, S, and R are not specified correctly, this procedure will result in the creation of artificial sources and/or sinks in the interior of the grid, and the resulting head configuration will be unrealistic.

Verification

Calibration means that, given a certain combination of parameters and boundary conditions, the model will produce field measured values of head at certain points in the grid. However, we have no guarantee that the combination of parameters found by trial and error is unique. Gillham and Farvolden (1974), for example, demonstrate that different hydraulic conductivity distributions will produce essentially the same head distribution. When better methods of solving the inverse problem are perfected, it may be possible to find an optimum solution for transmissivity.

The goal of verification is to demonstrate that the model is capable of simulating some historical hydrologic event for which field data are available. For example, one might attempt to simulate drawdowns during a pumping test or water level declines during a drought. Generally, some additional refinement of parameters will be necessary during verification. These refinements should be such that they do not change the steady-state calibration. After the model has been calibrated and verified, it is ready to be used for prediction. Karanjac et al. (1977) present a case study which illustrates the steps involved in model calibration, verification, and prediction.

Ideally, model calibration and verification should be done when the field investigation is underway so that field data required by the model can be readily collected. Furthermore, when used as a planning tool, a model should be updated and improved as new field data are collected.

Notes and Additional Reading

1. Fetter (1980, pp. 440–453) summarizes the main features of the Prickett–Lonnquist model. Prickett and Lonnquist (1971) present an application of their model to determine the effects of pumping in the Chicago area, and Karanjac et al. (1977) present an application to a water supply problem in the Uluova Plain, Turkey. Turk (1978) comments on this application and raises some objections which are discussed by Karanjac et al. in their reply. Land (1977) presents an interesting application of the Trescott et al. (1976) model to determine the hydraulic properties of a layered aquifer by using the results of an aquifer test. Winter (1976) uses the Trescott et al. model to simulate hypothetical groundwater-lake systems at steady-state. Mercer and Faust (1980c) discuss calibration and prediction procedures used in several groundwater models.

2. Trescott et al. (1976) present three options for solving their numerical model. One of these is iterative ADI. The other two methods are line successive over relaxation (LSOR) and the strongly implicit procedure (SIP). LSOR is similar to iterative ADI, but, instead of alternating directions from one iteration level to the next, the solution is oriented either along rows or along columns for the duration of the simulation. Whether one chooses to orient the solution along rows or columns depends on the relative magnitudes of the coefficients in the coefficient matrix. A relaxation factor is used to accelerate convergence.

   It has been demonstrated that SIP converges faster than either ADI or LSOR for problems involving heterogeneous or anisotropic media (Trescott et al., 1976). The strongly implicit procedure was introduced by Stone (1968). SIP is more implicit than ADI or LSOR because it is more closely related to a direct method of solving the matrix equation.

   Briefly stated, the method consists of solving a set of equations similar to the set generated by writing the implicit finite difference equation for the two-dimensional flow equation. That is, instead of solving the equation $[A]\{x\} = \{f\}$, one solves $[A + B]\{x\} = \{f\}$, where the matrix $[A + B]$ is constructed to be very close to the matrix $[A]$. The matrix $[A + B]$ is also constructed such that it can be decomposed into upper and lower triangular matrices, which have the desirable feature that a simple algorithm, similar to the Thomas algorithm, can be written to solve the set of equations generated by using $[A + B]\{x\} = \{f\}$. Most rows in the coefficient matrix $[A]$ contain five coefficients, one for each point in the five-point star (Figure 2.4). However, most rows in the coefficient matrix $[A + B]$ contain seven points, the points in the five-point star plus two points located on a diagonal line through point (i, j). The entries in $[A + B]$ are defined such that the coefficients of the points not in the five-point star are minimized. Remson et al. (1971) and Trescott et al. (1976) present detailed explanations of the method.

3. Several three-dimensional, transient flow models have also been developed. Freeze (1971) discusses a three-dimensional model for unsaturated-saturated systems and uses LSOR to solve the equations. Trescott (1975) and Trescott and Larson (1976, 1977) develop a three-dimensional groundwater flow model using SIP. Fleck and

McDonald (1978) present an application of the Trescott three-dimensional model to a groundwater flow system in the vicinity of a wastewater disposal system, and Winter (1978) uses the Trescott model to simulate a hypothetical groundwater-lake system at steady state.

4. In Chapter 5, we discussed the use of matrix methods for solving the transient flow equation. However, it should be apparent that these same techniques can also be used to solve the steady-state flow equations. Bennett (1976), for example, discusses the use of iterative ADI to solve Laplace's equation.

Direct solution techniques can be competitive with iterative methods for solving steady-state problems. Larson (1978) describes a direct solution algorithm which can be used in place of iterative ADI, LSOR, or SIP in the model developed by Trescott et al. (1976). This direct solution technique (known as the 4D ordering scheme with Gauss–Doolittle decomposition) can also be used to solve transient problems, but, for most transient problems, iterative schemes prove to be more efficient.

Problems

5.1 Run the computer program in Figure 5.2 and compare the results with the solution in Figure 4.3 and with the results of the analytical solution given in Problem 4.1.

5.2 Modify the computer program in Figure 5.2 to solve Equation 4.22, the one-dimensional transient equation for flow through an unconfined aquifer receiving recharge. Use the modified program to solve Problem 4.4.

CHAPTER 6

Finite Elements: Steady-State Flow

6.1 INTRODUCTION

The application of the finite element method to groundwater problems is a relatively recent development compared with the finite difference method. Each method leads to a set of algebraic equations in which the unknowns are the heads at a finite number of nodal points. In Figure 1.1 of Chapter 1, we showed the conceptual view of a problem domain as approximated by the two methods. The finite difference method is usually implemented with rectangular cells. The finite element method is implemented with a variety of element types, but the triangular element is a good beginning point for describing the method. Triangular elements are defined by three nodes—one at each corner. These nodes serve the purpose of locating unknown heads; that is, they are the points within the problem domain at which the heads are computed. Furthermore, the head within each element is defined in terms of the nodal values by basis or interpolation functions. The head is defined throughout the problem domain in a piecewise fashion over the individual elements. The use of interpolation functions to define the potential throughout the problem domain is an important concept that distinguishes the finite element method from the finite difference method. In the finite difference method, the head is defined only at the nodal points themselves. The definition of the head throughout the problem domain in the finite element method permits the application of variational or weighted residual principles.

Proponents of the finite element method point to its flexibility for problems in which the boundaries are irregular or for problems in which the medium is heterogeneous or anisotropic. Finite difference programs can also account for these complications. However, the flexibility of the finite element method is useful in solving coupled problems, such as contaminant transport, or in solving moving boundary problems, such as a moving water table. In the end, the choice of which method to use depends on such factors as the complexity of the problem and the user's familiarity with each method. For additional discussion of the advantages and disadvantages of each method, see Faust and Mercer (1980a).

Our purpose in this chapter is to describe the basic theory behind the finite element method and to develop a simple computer program that embodies the theory. Example problems, including seepage through a dam, illustrate some of the capabilities of the finite element method.

6.2 GALERKIN'S METHOD

Galerkin's method and the finite element technique are combined so frequently in computer solutions of groundwater problems that the two have become practically synonymous. Galerkin's method is based on a particular weighted residual principle which turns out to be equivalent to a variational principle, if one exists for the problem under consideration.

The philosophy behind a variational principle is that a physical quantity, such as the rate of energy dissipation, be minimized over the problem domain. This rate can be expressed in terms of the potential (head) throughout the domain. If the potential is expressed in terms of its nodal values, the variational principle leads to algebraic equations. The use of the variational principle and its equivalence to the Galerkin method is described in Appendix B.

A weighted residual principle is expressed directly in terms of the governing partial differential equation without need to resort to a physical quantity. The residual at each point in the problem domain is a measure of the degree to which the head does not satisfy the governing equation. If a particular weighted average of the residual is forced to vanish, the nodal heads are obtained as the solution of a system of algebraic equations.

We now proceed through the details of applying Galerkin's method to Laplace's equation. The first step is to define an approximate or trial solution, $\hat{h}(x, y)$. It is expressed as a series summation; each term is a product of a nodal head h_L and an associated nodal basis function $N_L(x, y)$.

$$\hat{h}(x, y) = \sum_{L=1}^{NNODE} h_L N_L(x, y) \tag{6.1}$$

The subscript L indicates nodal number and $NNODE$ is the total number of nodes in the problem domain. Note that a single subscript designates nodal number.

The basis functions in Equation 6.1 are analogous to unit vectors. The trial solution is built up as a linear combination of the basis functions. The basis functions are also called interpolation functions because they define the trial solution throughout the problem domain in terms of the nodal heads.

The next step is to require a total of $NNODE$ conditions to determine the $NNODE$ values of h_L. In the Galerkin method, the $NNODE$ conditions are that the residuals of the governing equation weighted by each of the $NNODE$ basis functions be zero when integrated over the entire domain of the problem:

$$\iint_D \left(\frac{\partial^2 \hat{h}}{\partial x^2} + \frac{\partial^2 \hat{h}}{\partial y^2} \right) N_L(x, y) \, dx \, dy = 0 \tag{6.2}$$

where $L = 1, 2, \ldots, NNODE$, and D signifies that the integration is done over the entire problem domain.

The quantity in parentheses is the residual. If the trial solution $\hat{h}(x, y)$ were exact, Laplace's equation would be satisfied throughout the problem domain and the residual would be zero everywhere. The residual is a measure of the extent to which $\hat{h}(x, y)$ does not satisfy Laplace's equation. In Galerkin's method, the requirement imposed is that $NNODE$ weighted averages of the residual vanish; the basis functions $N_L(x, y)$ are the weighting functions.

The choice of $N_L(x, y)$ as the weighting functions may seem ad hoc. For example, one might require instead that the residual vanish over $NNODE$ small subdomains about each nodal point; that is,

$$\iint_D \left(\frac{\partial^2 \hat{h}}{\partial x^2} + \frac{\partial^2 \hat{h}}{\partial y^2} \right) W_L \, dx \, dy = 0 \tag{6.3}$$

where W_L is 1 in subdomain L and is 0 outside subdomain L, $L = 1, 2, \ldots,$ $NNODE$. Indeed, Equation 6.3 leads to a system of equations to solve Laplace's equation approximately. The method is known as the subdomain method. A reason for the more frequent use of Galerkin's method is that the system of equations represented by Equation 6.2 is the same as the system generated from a minimum dissipation principle (see Appendix B). Variational principles do not exist for all governing equations. In such situations, Galerkin's method or other weighted residual methods can still be applied.

Integration by Parts

The basis functions $N_L(x, y)$ are usually defined in a piecewise but continuous manner over the problem domain D. However, the first derivatives of N_L may not be continuous. Because $\hat{h}(x, y)$ is a linear combination of the N_L's, its second derivatives are not easily defined at step discontinuities of the first derivatives, thereby complicating the evaluation of the integral in Equation 6.2. If integration by parts is applied to Equation 6.2, then the order of the derivatives in the integrand can be reduced by one.

Because the one-dimensional formula for integration by parts is most familiar, we consider first its application to the one-dimensional form of Equation 6.2:

$$\int_a^b \frac{d^2 \hat{h}}{dx^2} N_L(x) \, dx = 0 \tag{6.4}$$

The formula for integration by parts is

$$\int_a^b u \, dv = -\int_a^b v \, du + uv \Big|_a^b \tag{6.5}$$

To apply this formula to Equation 6.4, we let $u = N_L$ and $v = d\hat{h}/dx$. Then,

$$\int_a^b \frac{d^2 \hat{h}}{dx^2} N_L(x) \, dx = -\int_a^b \frac{d\hat{h}}{dx} \frac{dN_L}{dx} \, dx + N_L \frac{d\hat{h}}{dx} \Big|_a^b \tag{6.6}$$

The second term on the right-hand side of Equation 6.6 is proportional to the flux through the boundary weighted by N_L at the boundary points.

The generalization of Equation 6.6 to two dimensions is

$$\iint_D \left(\frac{\partial^2 \hat{h}}{\partial x^2} + \frac{\partial^2 \hat{h}}{\partial y^2} \right) N_L \, dx \, dy = - \iint_D \left(\frac{\partial \hat{h}}{\partial x} \frac{\partial N_L}{\partial x} + \frac{\partial \hat{h}}{\partial y} \frac{\partial N_L}{\partial y} \right) dx \, dy$$

$$+ \int_\Gamma \left(\frac{\partial \hat{h}}{\partial x} n_x + \frac{\partial \hat{h}}{\partial y} n_y \right) N_L \, d\sigma \qquad (6.7)$$

where Γ is the boundary of D, σ is a generalized variable representing distance along the boundary in a counterclockwise sense, and n_x and n_y are the components of a unit vector outwardly normal to Γ. The second term on the right-hand side of Equation 6.7 is proportional to the normal flux through the boundary weighted by N_L on the boundary. Note that if the flux is zero on the boundary, the boundary integral term is zero. The integrands on the right side of Equation 6.7 contain only first derivatives, which simplifies the problem considerably. The reason for the integration by parts will be seen more clearly in the next section when Galerkin's method is applied for basis functions whose first derivatives have step discontinuities within the problem domain.

6.3 TRIANGULAR ELEMENTS

We now describe the implementation of Galerkin's method according to the finite element technique for linear triangular elements. Each triangle is defined by nodes at its corners. The crux of the finite element procedure is to define basis functions $N_L(x, y)$ in Equation 6.1 that interpolate the nodal values in a piecewise fashion over those elements that contain node L. Equation 6.2 then provides $NNODE$ equations to determine each nodal head h_L, where $L = 1, 2, \ldots, NNODE$.

Finite Element Mesh

We illustrate the construction of a finite element mesh with the example of Section 2.2. The pictorial division of the problem domain of the region near a well in both the finite difference and finite element approximations is shown in Figure 6.1. Exactly the same set of sixteen nodal points is used in both cases. The four interior nodes represent unknowns, and the twelve boundary nodes represent specified heads. In the finite difference scheme, each nodal point is labeled by a row index and a column index. In the finite element scheme, the problem domain is divided into eighteen triangles. Each triangle is given an element number and three node numbers, one for each corner. Note that the nodes are numbered consecutively and systematically, column by column. The systematic numbering of nodes is not a requirement, but it can greatly reduce memory storage needs.

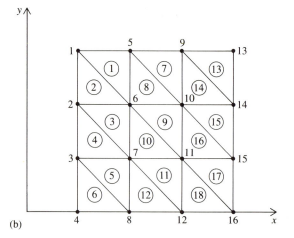

Figure 6.1
Finite difference and finite element representations of the region-near-a-well (Figure 2.3). The same nodal points are used for both methods.
(a) Finite difference mesh. The ordered pairs are column and row indices.
(b) Finite element mesh. Element numbers are circled. Node numbers are not circled.

The Archetypal Element

We need to describe those properties of a triangular element (Figure 6.2) which characterize such a finite element. Because each element is treated separately, and because nodes need not be located on a regular grid, we will no longer designate a node by the ordered pair (i, j) but by a single index number L and the nodal coordinates (x_L, y_L). The archetypal triangular element has node numbers i, j, and m in *counterclockwise* order. For example, element 3 in Figure 6.1b has node numbers $i = 2, j = 7$, and $m = 6$. The i, j, and m designations could have been cyclically permuted, that is, $i = 7, j = 6$, and $m = 2$ or $i = 6$, $j = 2$, and $m = 7$. The coordinates of nodes i, j, and m are designated as (x_i, y_i), (x_j, y_j), and (x_m, y_m), respectively. The unknowns of the problem are the heads at the nodes: $h_i = h(x_i, y_i)$, $h_j = h(x_j, y_j)$, and $h_m = h(x_m, y_m)$.

We define the trial solution $\hat{h}(x, y)$ throughout the triangular element by *linearly interpolating* the nodal values h_i, h_j, and h_m. A linear interpolation means that within the triangular element e,

$$\hat{h}^e(x, y) = a_0 + a_1 x + a_2 y \tag{6.8}$$

where a_0, a_1, and a_2 are coefficients that need to be determined. (Note that the symbol e is being used to designate element number. In this context, it does not represent the base of natural logarithms.) The coefficients can be determined by setting up three equations which require that the nodal values are obtained at the nodal coordinates.

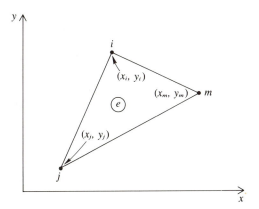

Figure 6.2
The archetypal triangular element e.
Nodes are labeled i, j, and m in counter-clockwise order.

$$h_i = a_0 + a_1 x_i + a_2 y_i \tag{6.9a}$$

$$h_j = a_0 + a_1 x_j + a_2 y_j \tag{6.9b}$$

$$h_m = a_0 + a_1 x_m + a_2 y_m \tag{6.9c}$$

If these equations are solved for a_0, a_1, and a_2, and if the expressions are substituted back into Equation 6.8, then Equation 6.8 can be rewritten as

$$\hat{h}^e(x, y) = N_i^e(x, y)h_i + N_j^e(x, y)h_j + N_m^e(x, y)h_m \tag{6.10}$$

where

$$N_i^e(x, y) = \frac{1}{2A^e} \left[(x_j y_m - x_m y_j) + (y_j - y_m)x + (x_m - x_j)y \right] \tag{6.11a}$$

$$N_j^e(x, y) = \frac{1}{2A^e} \left[(x_m y_i - x_i y_m) + (y_m - y_i)x + (x_i - x_m)y \right] \tag{6.11b}$$

$$N_m^e(x, y) = \frac{1}{2A^e} \left[(x_i y_j - x_j y_i) + (y_i - y_j)x + (x_j - x_i)y \right] \tag{6.11c}$$

and

$$2A^e = (x_i y_j - x_j y_i) + (x_m y_i - x_i y_m) + (x_j y_m - x_m y_j) \tag{6.12}$$

The A^e in Equation 6.12 is the area of triangle ijm expressed in terms of its coordinates at the corners. The functions $N_i^e(x, y)$, $N_j^e(x, y)$, and $N_m^e(x, y)$ are the element interpolation or basis functions. They are functions of the spatial coordinates x and y, and they define $\hat{h}(x, y)$ in element e in terms of the nodal values h_i, h_j, and h_m.

Although the algebraic expressions for the element basis functions $N_L^e(x, y)$ may seem lengthy, they are intuitively simple if a short list of their properties is kept in mind.

1. N_L^e is 1 at node L and 0 at the other two nodes.
2. N_L^e varies linearly with distance along any side.
3. N_L^e is $\frac{1}{3}$ at the centroid of the triangle.
4. N_L^e is 0 along the side opposite node L.

Patch of Elements

The collection of elements which contain a specific node L forms a patch around node L (Figure 6.3a). Within the patch, the nodal basis function $N_L(x, y)$ is defined piecewise by the element basis functions $N_L^e(x, y)$ over each element

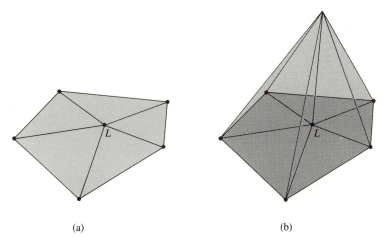

(a)   (b)

Figure 6.3
Definition of nodal basis function $N_L(x, y)$ over patch of elements containing node L.
(a) Plan view of patch of elements about node L.
(b) Three-dimensional view of nodal basis function $N_L(x, y)$.

belonging to the patch. Outside the patch, $N_L(x, y)$ is 0. Thus, each basis function $N_L(x, y)$ is pyramidal in shape (Figure 6.3b). Its peak value of 1 is located directly over node L. Its edges slope to 0 at all the other nodes.

6.4 ASSEMBLY OF CONDUCTANCE MATRIX

A system of algebraic equations is represented by Equation 6.2. We could build up sequentially the system of equations one nodal number at a time. A method more efficient than considering a single row at a time is to proceed through the problem sequentially by elements and to incorporate the contributions of element ijm to the three rows $L = i$, $L = j$, and $L = m$. After the contributions of the last element have been summed, the total system of equations is completed.

The element-by-element definition of N_L means that, although it is piecewise continuous, its first derivatives contain step discontinuities over element boundaries. If we use the integration by parts result, Equation 6.7, then Equation 6.2 can be written as the summation of integrals over individual elements being equal to a boundary integral.

$$\sum_e \left\{ \iint_e \left(\frac{\partial \hat{h}^e}{\partial x} \frac{\partial N_L}{\partial x} + \frac{\partial \hat{h}^e}{\partial y} \frac{\partial N_L}{\partial y} \right) dx\, dy \right\} = \int_\Gamma \left(\frac{\partial \hat{h}}{\partial x} n_x + \frac{\partial \hat{h}}{\partial y} n_y \right) N_L\, d\sigma \quad (6.13)$$

where $L = 1, 2, \ldots, NNODE$. Note that on the left-hand side of Equation 6.13, $\hat{h}$ has been replaced by $\hat{h}^e$.

Equation 6.10, which defines $\hat{h}^e(x, y)$, shows that $\partial \hat{h}^e / \partial x$ and $\partial \hat{h}^e / \partial y$ can be written in terms of the nodal heads h_i, h_j, and h_m and derivatives of the element basis functions:

$$\frac{\partial \hat{h}^e}{\partial x} = \frac{\partial N_i^e}{\partial x} h_i + \frac{\partial N_j^e}{\partial x} h_j + \frac{\partial N_m^e}{\partial x} h_m \quad (6.14a)$$

$$\frac{\partial \hat{h}^e}{\partial y} = \frac{\partial N_i^e}{\partial y} h_i + \frac{\partial N_j^e}{\partial y} h_j + \frac{\partial N_m^e}{\partial y} h_m \quad (6.14b)$$

The derivatives of N_L^e can be written in terms of the nodal coordinates by differentiation of Equation 6.11. If for each element Equation 6.14 is substituted back into the left-hand side of Equation 6.13, then Equation 6.13 can be seen to have the form

$$G_{L,1}h_1 + \cdots + G_{L,i}h_i + \cdots + G_{L,j}h_j$$
$$+ \cdots + G_{L,m}h_m + \cdots + G_{L,NNODE}h_{NNODE} = f_L \quad (6.15)$$

where $L = 1, 2, \ldots, NNODE$.

Explicit expressions for the contribution from element e to the coefficients $G_{L,i}$, $G_{L,j}$, and $G_{L,m}$ are derived in the following subsection. Expressions for f_L are given in the section on boundary conditions.

The system of $NNODE$ equations represented by Equation 6.15 can be written in matrix form as

$$[G]\{h\} = \{f\} \quad (6.16)$$

The $[G]$ matrix is a square coefficient matrix called the conductance matrix; the $\{h\}$ matrix is a column matrix of nodal heads h_L; the $\{f\}$ matrix is a column matrix which represents the boundary conditions. Thus the Galerkin method will lead to a set of algebraic equations which can be put in the matrix notation described in Section 5.2.

The reason $[G]$ is called the conductance matrix is not so obvious for Laplace's equation. If the hydraulic conductivity were left in the governing equation, then it would appear as a multiplicative factor in Equation 6.13 and in entries of $[G]$.

Element Conductance Matrix

The subdivision of the problem domain into elements means that we evaluate the integrals in the left-hand side of Equation 6.13 one element at a time. A nodal basis function N_L is nonzero only over the patch of elements about node L. Therefore, the double integral over an element e is nonzero only if the element is in the patch about node L, in which case N_L is defined within the element to be the element basis function N_L^e. In other words, an element e made up of nodes i, j, and m contributes only to those three equations in which $L = i, j$, or m. Furthermore, in each of those three equations, element e contributes only to terms in h_i, h_j, and h_m. Altogether, element e contributes to three rows and three columns of the coefficient matrix $[G]$. The contribution of element e can be thought of as a three-by-three matrix

$$[G^e] = \begin{pmatrix} G_{ii}^e & G_{ij}^e & G_{im}^e \\ G_{ji}^e & G_{jj}^e & G_{jm}^e \\ G_{mi}^e & G_{mj}^e & G_{mm}^e \end{pmatrix} \tag{6.17}$$

where $[G^e]$ is called the element conductance matrix.

The individual terms of $[G^e]$ are found by evaluating the double integral over element e in Equation 6.13. The function N_L is replaced by N_L^e according to the preceding discussion.

$$\iint_e \left(\frac{\partial \hat{h}^e}{\partial x} \frac{\partial N_L^e}{\partial x} + \frac{\partial \hat{h}^e}{\partial y} \frac{\partial N_L^e}{\partial y} \right) dx\, dy = \iint_e \left\{ \left(\frac{\partial N_i^e}{\partial x} h_i + \frac{\partial N_j^e}{\partial x} h_j + \frac{\partial N_m^e}{\partial x} h_m \right) \frac{\partial N_L^e}{\partial x} \right.$$

$$\left. + \left(\frac{\partial N_i^e}{\partial y} h_i + \frac{\partial N_j^e}{\partial y} h_j + \frac{\partial N_m^e}{\partial y} h_m \right) \frac{\partial N_L^e}{\partial y} \right\} dx\, dy$$

$$\tag{6.18}$$

where $L = i, j$, or m.

The integrand, which involves spatial first derivatives of the basis functions, is independent of x and y because the basis functions are linear in x and y. In this case, the result of the integration is the integrand multiplied by the area A^e of the element. If like coefficients of h_i, h_j, and h_m are gathered, then

$$\iint_e \left(\frac{\partial \hat{h}^e}{\partial x}\frac{\partial N_L^e}{\partial x} + \frac{\partial \hat{h}^e}{\partial y}\frac{\partial N_L^e}{\partial y}\right) dx\, dy = A^e\left(\frac{\partial N_i^e}{\partial x}\frac{\partial N_L^e}{\partial x} + \frac{\partial N_i^e}{\partial y}\frac{\partial N_L^e}{\partial y}\right)h_i$$

$$+ A^e\left(\frac{\partial N_j^e}{\partial x}\frac{\partial N_L^e}{\partial x} + \frac{\partial N_j^e}{\partial y}\frac{\partial N_L^e}{\partial y}\right)h_j$$

$$+ A^e\left(\frac{\partial N_m^e}{\partial x}\frac{\partial N_L^e}{\partial x} + \frac{\partial N_m^e}{\partial y}\frac{\partial N_L^e}{\partial y}\right)h_m \qquad (6.19)$$

where $L = i, j,$ or m. The coefficients of h_i, h_j, and h_m in Equation 6.19 are the column entries along row L of the element conductance matrix, that is,

$$G_{L,i}^e = A^e\left(\frac{\partial N_i^e}{\partial x}\frac{\partial N_L^e}{\partial x} + \frac{\partial N_i^e}{\partial y}\frac{\partial N_L^e}{\partial y}\right) \qquad (6.20a)$$

$$G_{L,j}^e = A^e\left(\frac{\partial N_j^e}{\partial x}\frac{\partial N_L^e}{\partial x} + \frac{\partial N_j^e}{\partial y}\frac{\partial N_L^e}{\partial y}\right) \qquad (6.20b)$$

$$G_{L,m}^e = A^e\left(\frac{\partial N_m^e}{\partial x}\frac{\partial N_L^e}{\partial x} + \frac{\partial N_m^e}{\partial y}\frac{\partial N_L^e}{\partial y}\right) \qquad (6.20c)$$

where $L = i, j,$ or m.

Global Conductance Matrix

The element conductance matrix represents element e's contribution to three rows and three columns of the matrix $[G]$ in Equation 6.16. The matrix $[G]$ is called the global conductance matrix. The summation over elements in Equation 6.13 means that the element contributions to the global conductance matrix are summed; that is,

$$G_{L,i} = \sum_e G_{L,i}^e \qquad (6.21)$$

for all L and i.

Note that the $[G]$ matrix is not constructed one row at a time. Rather, the element contributions to the global coefficient matrix are sorted out by the subscript pairs—L, i; L, j; and L, m—in Equation 6.20 as the elements are considered one at a time. The process is illustrated by Figure 6.4. For example, element number 1 is defined by nodes $i = 1, j = 2$, and $m = 3$. The nine values which make up the element matrix $[G^{e=1}]$ are dispersed according to their subscript numbers into the correct global matrix locations. These rows and columns are indicated by X's. This process is done for each of the six elements in turn. The summation

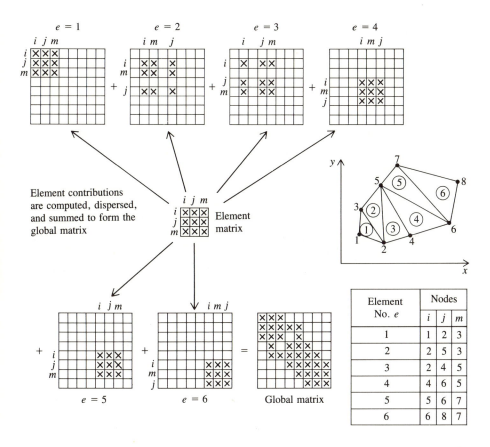

Figure 6.4
Schematic diagram showing the relationship of element conductance matrices and the global conductance matrix. Inset shows the finite element mesh under consideration. (After Cheng, 1978.)

of all the element contributions leads to the final configuration of entries in the global matrix.

The definition of the element matrix contributions, Equation 6.20, implies that the matrix is symmetric, that is $G_{L,i}^e = G_{i,L}^e$. Hence, the global matrix is also symmetric, that is, $G_{L,i} = G_{i,L}$ for all L and i.

The assembly of the global conductance matrix is done strictly in terms of the geometry—node and element labeling—of the finite element mesh. If the aquifer were not homogeneous or not isotropic, the global matrix would include the aquifer hydraulic conductivities.

6.5   BOUNDARY CONDITIONS

What remains to be incorporated are the boundary conditions, either specified flow or specified head. The right-hand side of Equation 6.13 is a boundary integral which is proportional to a weighted average of the normal flux. Yet, the presence of the boundary integral involving knowledge of the boundary flow may seem contradictory for a boundary along which heads are specified.

For the Lth equation, the Lth nodal basis function N_L is the weighting function. Therefore, the components of the column vector $\{f\}$ in Equation 6.16 are given by the boundary integral in Equation 6.13, one for each nodal basis function N_L. First, consider L to be an interior node. If no side of an element in the patch about L is part of the boundary, then $N_L = 0$ over the whole boundary. Even if a side of an element in the patch is part of the boundary, it will be a side opposite L, and $N_L = 0$ along this boundary side by the fourth property of N_L^e described in Section 6.3. Because N_L is the weighting function of the boundary integral in Equation 6.13, the entire integral must be zero for all interior nodes L. Hence, $f_L = 0$ if the subscript L represents an interior node, regardless of boundary conditions. We next treat separately the handling of boundary conditions when L represents a boundary node.

Specified Flow

Let L be a boundary node on a boundary across which the normal flux is specified (Figure 6.5). The boundary integral in Equation 6.13 can be written in terms of the specified flux by applying Darcy's law.

$$\int_\Gamma \left(\frac{\partial \hat{h}}{\partial x} n_x + \frac{\partial \hat{h}}{\partial y} n_y \right) N_L \, d\sigma = \int_i^L \frac{q_1}{K} N_L(\sigma) \, d\sigma + \int_L^m \frac{q_2}{K} N_L(\sigma) \, d\sigma \quad (6.22)$$

where K is the hydraulic conductivity, and σ is a generalized variable representing distance along the boundary. The sign convention is that q_1 and q_2 are positive when they represent inflow through the boundary. The boundary integral is nonzero only over the two line segments iL and Lm because $N_L = 0$ for the boundary beyond nodes i and m. The interpolation function N_L varies linearly from 1 to 0 between nodes L and i and between nodes L and m. Hence, the integrals can be done by inspection to give the Lth entry of the $\{f\}$ vector to be

$$f_L = \frac{q_1}{K}\frac{\overline{iL}}{2} + \frac{q_2}{K}\frac{\overline{Lm}}{2} \tag{6.23}$$

where $\overline{iL}$ is the distance between nodes i and L, and $\overline{Lm}$ is the distance between nodes L and m. The volumetric flow through a side is distributed equally to the two nodes which make up the side.

In summary, flow boundary conditions are incorporated into the column vector $\{f\}$ of Equation 6.16. For all interior nodes or nodes on a no-flow boundary, $f_L = 0$. For boundary nodes on a specified flow boundary, f_L is given by Equation 6.23.

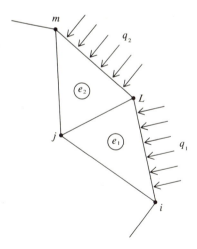

Figure 6.5
Finite element treatment of a
specified flow boundary condition.

Specified Head

The case we have yet to consider is the value of the boundary integral in Equation 6.13 for a boundary node L for which the nodal head h_L is known. If h_L is known, then the Lth equation must be redundant. The boundary integral value f_L is immaterial if we simply eliminate the Lth equation from consideration. At the same time, we use the boundary value to transfer terms containing h_L in the remaining rows of $[G]\{h\}$ into the $\{f\}$ matrix. If the complete system of equations represented by Equation 6.16 is solved by iteration, we can simply skip over those equations L in which h_L is a known boundary value. Because the known value is used as the initial guess, and because it is never changed, the specified value of h_L is used whenever it appears in other equations. This method of handling specified boundary heads is applied in the computer program in Section 6.6.

If a direct matrix solution technique is used, a numerical trick can handle specified heads h_L without eliminating equations. The diagonal component $G_{L,L}$ of the global coefficient matrix is multiplied by 10^{10}, and f_L is set equal to $10^{10}G_{L,L}h_L$, where the specified value of h_L is used. The Lth equation will be

$$G_{L,1}h_1 + G_{L,2}h_2 + \cdots + 10^{10}G_{L,L}h_L + \cdots + G_{L,n}h_n = 10^{10}G_{L,L}h_L \quad (6.24)$$

When the complete system of equations is solved for the column vector $\{h\}$, h_L is forced to its specified value because all the other terms in the Lth equation are insignificant relative to the diagonal term.

In summary, fixed boundary conditions lead to a reduction in the number of unknowns. The value of f_L for a boundary node L is irrelevant and can be set to zero. Because the value of $f_L = 0$ for all interior nodes, a problem whose boundaries are all specified has $\{f\} = 0$.

6.6   FINITE ELEMENT COMPUTER PROGRAM

The concepts and equations of Sections 6.3, 6.4, and 6.5 are implemented by the computer program shown in Figure 6.6. The execution of the program is illustrated by the Dirichlet boundary problem described in Section 2.2.

The program is completely self-contained and requires no subroutines. It illustrates the general architecture of all finite element programs, but it is not efficient in its use of memory or in its speed of operation. The program can be divided into three major blocks.

Block 1.   Obtain nodal coordinates. Specify boundary nodes with fixed heads, and specify initial guesses for interior nodes (lines 6 to 21).

Block 2.   Construct the global conductance matrix by summing the contributions from each element (lines 23 to 50).

Block 3.   Solve the system of linear equations by iteration (lines 52 to 73).

Node Specification (Lines 6 to 21)

Block 1 of the program contains two *READ* statements. Each *READ* is followed by a *PRINT* to echo the input cards. The first *READ* (line 11) obtains the number of nodes *NNODE* and number of elements *NELEM*. The second *READ* (line 16) is in a *DO* loop which obtains for each node the (X, Y) coordinates, the node type, and the boundary value or initial guess. The input cards containing the nodal information are assumed to be in numerical sequence. That is, the coordinates of node 1 are on the first card, the coordinates of node 2 are on the second card, and so on. The coordinates are stored in one-dimensional arrays $X(L)$ and $Y(L)$, where L is the node number. The value of the flag $KODE(L)$ is one for boundary nodes and is zero for interior nodes. If $KODE(L) = 1$, then the value of $H(L)$ read in should be the specified value for boundary node L. If $KODE(L) = 0$, then the value of $H(L)$ read in is the initial guess for interior node L.

Global Conductance Matrix (Lines 23 to 50)

Block 2, the construction of the conductance matrix $[G]$, is the central portion of the finite element program. The summation over all elements is represented by line 25, *DO* 100 $K = 1,NELEM$, where the index K replaces the designation e for element number. The node numbers of the Kth element are obtained in line 27, *READ* 31,*I,J,M*. Thus, the input cards must contain one element's node numbers per card, and they must be in numerical sequence. That is, the node numbers of element 1 must be the first card of the set, the node numbers of element 2 must be the second card of the set, and so on.

Figure 6.6
Finite element computer program to solve Laplace's equation.

```
1.      C    FINITE ELEMENT PROGRAM FOR LAPLACE'S EQUATION.   TRIANGULAR ELEMENTS.
2.           REAL NX,NY
3.           DIMENSION X(25),Y(25),H(25),G(25,25),KODE(25)
4.           DIMENSION NX(3),NY(3),NODE(3)
5.      C
6.      C    BLOCK 1.   OBTAIN NODAL COORDINATES AND INITIALIZE ARRAYS.
7.      C
8.           PRINT 1
9.      1    FORMAT(1H1,'ECHO OF INPUT CARDS',//)
10.     C    NUMBER OF NODES AND ELEMENTS.
11.          READ 5,NNODE,NELEM
12.          PRINT 5,NNODE,NELEM
13.     5    FORMAT(2I10)
14.          DO 10 L=1,NNODE
15.     C    NODAL COORDINATES, TYPE, AND SPECIFIED OR DEFAULT VALUE.
16.          READ 11,X(L),Y(L),KODE(L),H(L)
17.     10   PRINT 11,X(L),Y(L),KODE(L),H(L)
18.     11   FORMAT(2F10.2,I10,F10.2)
19.          DO 20 L=1,NNODE
20.          DO 20 JJ=1,NNODE
21.     20   G(L,JJ)=0.
22.     C
```

```
23.     C   BLOCK 2.   CONSTRUCT THE CONDUCTANCE MATRIX.
24.     C
25.           DO 100 K=1,NELEM
26.     C   NODE NUMBERS OF ELEMENT K.
27.           READ 31,I,J,M
28.           PRINT 31,I,J,M
29.     31    FORMAT(3I10)
30.           A=0.5*((X(I)*Y(J)-X(J)*Y(I))+(X(M)*Y(I)-X(I)*Y(M))
31.          1       +(X(J)*Y(M)-X(M)*Y(J)))
32.     C   NX AND NY ARE SPATIAL DERIVATIVES OF INTERPOLATION FUNCTIONS.
33.           NX(1)=0.5*(Y(J)-Y(M))/A
34.           NX(2)=0.5*(Y(M)-Y(I))/A
35.           NX(3)=0.5*(Y(I)-Y(J))/A
36.           NY(1)=0.5*(X(M)-X(J))/A
37.           NY(2)=0.5*(X(I)-X(M))/A
38.           NY(3)=0.5*(X(J)-X(I))/A
39.           NODE(1)=I
40.           NODE(2)=J
41.           NODE(3)=M
42.           DO 40 KK=1,3
43.           L=NODE(KK)
44.           G(L,I)=G(L,I)+A*(NX(1)*NX(KK)+NY(1)*NY(KK))
45.           G(L,J)=G(L,J)+A*(NX(2)*NX(KK)+NY(2)*NY(KK))
46.           G(L,M)=G(L,M)+A*(NX(3)*NX(KK)+NY(3)*NY(KK))
47.     40    CONTINUE
48.     100   CONTINUE
```

(*Continued*)

[131]

Figure 6.6 (Continued)

```
49.       PRINT 999,((G(I,J),I=1,NNODE),J=1,NNODE)
50. 999   FORMAT(1H1,'CONDUCTANCE MATRIX'//,16(16F5.1//))
51. C
52. C     BLOCK 3.   SOLVE SYSTEM OF EQUATIONS BY ITERATION.
53. C
54. 200   AMAX=0.
55.       DO 400 L=1,NNODE
56. C     EXCLUDE FIXED BOUNDARY HEADS FROM ITERATION.
57.       IF(KODE(L).EQ.1)GO TO 400
58.       OLDVAL=H(L)
59.       SUM=0.
60.       DO 300 JJ=1,NNODE
61.       IF(JJ.EQ.L)GO TO 300
62.       SUM=SUM+G(L,JJ)*H(JJ)
63. 300   CONTINUE
64.       H(L)=-SUM/G(L,L)
65.       ERR=ABS(OLDVAL-H(L))
66.       IF(ERR.GT.AMAX)AMAX=ERR
67. 400   CONTINUE
68.       IF(AMAX.GT.0.01)GO TO 200
69.       PRINT 405
70. 405   FORMAT(////,1X,'NODE NUMBER',6X,'HEAD')
71.       DO 410 L=1,NNODE
72.       PRINT 411,L,H(L)
73. 411   FORMAT(I7,5X,F10.2)
74.       STOP
75.       END
```

The archetypal triangle IJM contributes to rows I, J, and M and to columns I, J, and M of the $[G]$ matrix according to Equation 6.20. The program cycles through the cases $L = I$, $L = J$, and $L = M$ in the loop DO 40 $KK = 1,3$ (line 42). For each case, the element contributions to the terms $G(L, I)$, $G(L, J)$, and $G(L, M)$ of the global conductance matrix are computed and summed (lines 44 to 46). The partial derivatives of the interpolation functions required by Equation 6.20 are computed in lines 33 to 38. The program notation is that NX is a partial derivative with respect to x, and NY is a partial derivative with respect to y. The subscripts 1, 2, and 3 correspond to subscripts i, j, and m.

The column matrix $\{f\} = 0$ for the case that all boundary heads are specified. The system of linear equations $[G]\{h\} = 0$, which represents this case, has been constructed after all the element contributions have been summed.

Iterative Solution of Equations (Lines 52 to 73)

The solution of $[G]\{h\} = 0$ could be obtained by a direct matrix solution technique. However, we use iteration to make this finite element program entirely self-contained and short. The Gauss–Seidel iteration used is the same as that used in the finite difference programs of Chapter 2. The Lth linear equation is

$$\sum_{JJ=1}^{NNODE} G_{L,JJ} h_{JJ} = 0 \tag{6.25}$$

where JJ is the name of the summation index. Equation 6.25 can be solved for h_L:

$$h_L = -\frac{1}{G_{L,L}} \left(\sum_{\substack{JJ=1 \\ JJ \neq L}}^{NNODE} G_{L,JJ} h_{JJ} \right) \tag{6.26}$$

The notation means that the summation is for all $JJ = 1$ to $JJ = NNODE$, except for $JJ = L$. The Gauss–Seidel formula for h_L is applied in lines 59 to 64 of the computer program. The variable SUM represents the summation quantity in Equation 6.26.

A single iteration consists of applying the Gauss–Seidel formula from $L = 1$ to $L = NNODE$. However, fixed boundary heads are excluded (line 57) from the iteration. A solution is reached when successive iterations lead to a maximum change in any h_L which is below a set tolerance level.

6.7 REGION-NEAR-A-WELL EXAMPLE

The finite element program of Section 6.6 is used to solve the Dirichlet boundary problem of Section 2.2. The echo of the input cards is shown in Figure 6.7 for the node and element numbering scheme of Figure 6.1b. The first card contains the number of nodes $NNODE$ and number of elements $NELEM$. It is followed successively by the nodal information and the node numbers of the elements.

The printout of the conductance matrix is shown in Figure 6.8. Because the problem contains 16 nodes, $[G]$ contains 16 rows and 16 columns. As an example of the linear equations generated, the sixth row of $[G]$ provides the equation

$$-h_2 - h_5 + 4h_6 - h_7 - h_{10} = 0 \qquad (6.27)$$

The solution for h_6 in Gauss–Seidel iteration is

$$h_6 = \frac{h_2 + h_5 + h_7 + h_{10}}{4} \qquad (6.28)$$

Note that in Figure 6.1b, node 6 is at the center of the five-point star formed by nodes 2, 5, 6, 7, and 10. Thus the equations for the interior nodes by the finite element method in this case are just the same as the usual five-point star (Figure 2.4) from the finite difference method. The nodal heads are also printed in Figure 6.8, and they may be compared with those obtained in Chapter 2 (Figure 2.8).

Irregular Mesh

The regular mesh of Figure 6.1b is not a requirement of the finite element method. The Picasso-style mesh (Figure 6.9) for the region-near-a-well contains the same nodal points as before, but it has a different element configuration. This mesh and the computer program shown in Figure 6.6 can be used to solve the problem for the unknown heads h_6, h_7, h_{10}, and h_{11}.

The conductance matrices generated for the regular mesh and for the Picasso mesh are symmetric; that is, entries reflect across the diagonal. However, the Picasso conductance matrix leads to a system of linear equations that is not the same as the usual set of finite difference equations.

The bandwidth of a symmetric matrix is the maximum number of columns between the diagonal and the last nonzero entry, inclusive, along any row. The

ECHO OF INPUT CARDS

16	18		
100.00	300.00	1	8.04
100.00	200.00	1	7.68
100.00	100.00	1	7.19
100.00	.00	1	6.82
200.00	300.00	1	8.18
200.00	200.00	0	.00
200.00	100.00	0	.00
200.00	.00	1	7.56
300.00	300.00	1	8.36
300.00	200.00	0	.00
300.00	100.00	0	.00
300.00	.00	1	7.99
400.00	300.00	1	8.53
400.00	200.00	1	8.41
400.00	100.00	1	8.33
400.00	.00	1	8.29
1	6	5	
1	2	6	
2	7	6	
2	3	7	
3	8	7	
3	4	8	
5	10	9	
5	6	10	
6	11	10	
6	7	11	
7	12	11	
7	8	12	
9	14	13	
9	10	14	
10	15	14	
10	11	15	
11	16	15	
11	12	16	

Figure 6.7
Echo of input cards for region-near-a-well problem (Figure 2.3).

(a) CONDUCTANCE MATRIX

1.0	-.5	.0	.0	-.5	.0	.0	.0	.0	.0	.0	.0	.0	.0	.0	.0
-.5	2.0	-.5	.0	.0	-1.0	.0	.0	.0	.0	.0	.0	.0	.0	.0	.0
.0	-.5	2.0	-.5	.0	.0	-1.0	.0	.0	.0	.0	.0	.0	.0	.0	.0
.0	.0	-.5	1.0	.0	.0	.0	-.5	.0	.0	.0	.0	.0	.0	.0	.0
-.5	.0	.0	.0	2.0	-1.0	.0	.0	-.5	.0	.0	.0	.0	.0	.0	.0
.0	-1.0	.0	.0	-1.0	4.0	-1.0	.0	.0	-1.0	.0	.0	.0	.0	.0	.0
.0	.0	-1.0	.0	.0	-1.0	4.0	-1.0	.0	.0	-1.0	.0	.0	.0	.0	.0
.0	.0	.0	-.5	.0	.0	-1.0	2.0	.0	.0	.0	-.5	.0	.0	.0	.0
.0	.0	.0	.0	-.5	.0	.0	.0	2.0	-1.0	.0	.0	-.5	.0	.0	.0
.0	.0	.0	.0	.0	-1.0	.0	.0	-1.0	4.0	-1.0	.0	.0	-1.0	.0	.0
.0	.0	.0	.0	.0	.0	-1.0	.0	.0	-1.0	4.0	-1.0	.0	.0	-1.0	.0
.0	.0	.0	.0	.0	.0	.0	-.5	.0	.0	-1.0	2.0	.0	.0	.0	-.5
.0	.0	.0	.0	.0	.0	.0	.0	-.5	.0	.0	.0	1.0	-.5	.0	.0
.0	.0	.0	.0	.0	.0	.0	.0	.0	-1.0	.0	.0	-.5	2.0	-.5	.0
.0	.0	.0	.0	.0	.0	.0	.0	.0	.0	-1.0	.0	.0	-.5	2.0	-.5
.0	.0	.0	.0	.0	.0	.0	.0	.0	.0	.0	-.5	.0	.0	-.5	1.0

(b) NODE NUMBER

NODE NUMBER	HEAD
1	8.04
2	7.68
3	7.19
4	6.82
5	8.18
6	7.93
7	7.68
8	7.56
9	8.36
10	8.19
11	8.05
12	7.99
13	8.53
14	8.41
15	8.33
16	8.29

Figure 6.8
(a) Global conductance matrix and (b) final nodal heads for region-near-a-well problem.

[137]

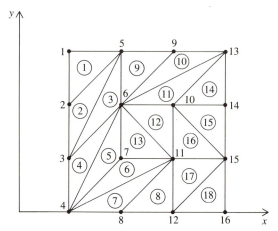

Figure 6.9
Picasso-style finite element mesh for region-near-a-well problem.

bandwidth of the conductance matrix in Figure 6.8 is five, whereas the bandwidth for the Picasso conductance matrix is eight. Systematic node and element numbering leads to smaller bandwidths of the conductance matrix, thereby requiring less computer storage. Also, more accurate answers are obtained when large, obtuse angles are avoided when constructing elements.

6.8 SEEPAGE THROUGH A DAM

We used Dupuit assumptions to find the height of the water table in an earthen dam (Figure 3.8). Because of the Dupuit assumption that flow through the dam be horizontal, the problem was reduced to one dimension. In this section, we maintain the exact two-dimensional formulation of the seepage problem and use the finite element method for solution.

The cross section of the 6-m wide dam is shown in Figure 6.10. The triangular mesh contains sixteen nodes and eighteen elements just as in the mesh for the region-near-a-well (Figure 6.1b). The top boundary must satisfy two constraints. (a) The surface is a no-flow boundary. (b) The head at each point on the boundary must equal its elevation. The pressure at each point on the boundary is zero because there is no overlying column of water at the top of the saturated zone. Although we have two constraints that must be satisfied on a single boundary, we have the degree of freedom of locating the boundary. The unknowns of the problem are the y coordinates $y_5 = h_5$, $y_9 = h_9$, and $y_{13} = h_{13}$ and the interior heads h_6, h_7, h_{10}, and h_{11}.

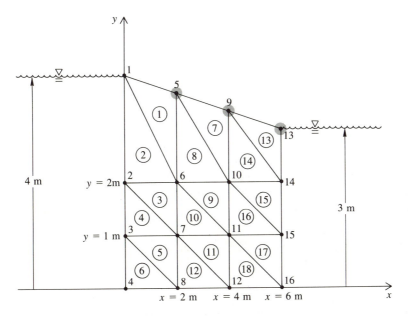

Figure 6.10
Finite element mesh for the two-dimensional seepage through a dam problem. The dam itself is not shown in the figure. Element numbers are circled. Node numbers are not circled.

Self-Consistent Iterative Solution

The three unknown y coordinates are solved for iteratively. The Dupuit assumption values (Figure 3.11), $y_5 = 3.70$ m, $y_9 = 3.36$ m, and $y_{13} = 3.00$ m, can serve as the initial guesses. The top boundary must satisfy the constraint that it be a no-flow boundary. The bottom boundary is also a no-flow boundary under the assumption that the dam rests on impermeable bedrock. Specified head boundary conditions apply for side nodes: $h_1 = h_2 = h_3 = h_4 = 4.00$ m, and $h_{14} = h_{15} = h_{16} = 3.00$ m. The governing equaton is Laplace's equation. The boundary value problem is thus completely defined and can be solved for heads by the finite element computer program.

After an iteration, the boundary heads h_5, h_9, and h_{13} computed by the program are compared against the assumed values for y_5, y_9, and y_{13}. If $h_5 \simeq y_5$, $h_9 \simeq y_9$, and $h_{13} \simeq y_{13}$ within a preset tolerance, then the second constraint that the top boundary heads equal elevation heads is satisfied, and we have a solution. If the discrepancies are too large, we begin another iteration with $y_5 = h_5$, $y_9 = h_9$, and $y_{13} = h_{13}$, where the heads are those just computed in the previous step. When a seepage face exists—that is, $y_{13} > 3.00$ m—then the seepage outflow between nodes 13 and 14 is a specified flow boundary. In the iterative procedure, the seepage outflow value is calculated by using the head values determined in the previous step. At each iteration step, we are actually changing the problem domain and seepage face boundary condition as we re-locate the top boundary to be consistent with the heads computed from the previous location of the boundary. The finite element method, with its flexi-bility for incorporating the changing location of nodal coordinates (that is, deformable elements), enables straightforward solution of the boundary value problem at each iteration.

Computer Program

The computer program is shown in Figure 6.11, the echo of the input cards is shown in Figure 6.12, and the answers are shown in Figure 6.13. The major change to the program is the addition of BLOCK 4 (lines 76 to 94), where the consistency tests $H(5) = Y(5)$, $H(9) = Y(9)$, and $H(13) = Y(13)$ are made. Other changes are in BLOCK 2. The node numbers for element K are now stored in an array where $I = NODE(K, 1)$, $J = NODE(K, 2)$, and $M = NODE(K, 3)$ be-cause, with each iteration, the node numbers of each element are required in order to construct the updated conductance matrix. The printing of the con-ductance matrix has been removed. The seepage flow normal to the right face of element 13 is computed as QX in line 47 in accordance with Equation 6.22. The flow boundary condition is incorporated into the $\{f\}$ matrix for $L = 13$ and $L = 14$.

The final water table elevations are $Y(5) = 3.72$ m, $Y(9) = 3.40$ m, and $Y(13) = 3.08$ m. Note that $Y(13)$ is greater than 3.00 m, and hence a seepage face of 0.08 m exists. The seepage face cannot exist under Dupuit assumptions because $Y(13) = H(13) = 3.00$ m is a boundary condition in that case.

Figure 6.11

Finite element computer program to solve for location of free surface boundary and heads in dam.

```
1.    C    FINITE ELEMENT MODEL FOR FREE SURFACE.
2.         REAL NX,NY
3.         DIMENSION X(25),Y(25),H(25),G(25,25),F(25),KODE(25)
4.         DIMENSION NX(3),NY(3),NODE(25,3)
5.         NLOOP=0
6.    C
7.    C    BLOCK 1.  OBTAIN NODAL COORDINATES AND INITIALIZE ARRAYS.
8.    C
9.         PRINT 1
10.   1     FORMAT(1H1,'ECHO OF INPUT CARDS',//)
11.   C    NUMBER OF NODES AND ELEMENTS.
12.        READ 5,NNODE,NELEM
13.        PRINT 5,NNODE,NELEM
14.   5     FORMAT(2I10)
15.        DO 10 L=1,NNODE
16.   C    NODAL COORDINATES, TYPE, AND SPECIFIED OR DEFAULT VALUE.
17.        READ 11,X(L),Y(L),KODE(L),H(L)
18.   10    PRINT 11,X(L),Y(L),KODE(L),H(L)
19.   11    FORMAT(2F10.2,I10,F10.2)
20.        DO 20 L=1,NNODE
21.        F(L)=0.
22.        DO 20 JJ=1,NNODE
23.   20    G(L,JJ)=0.
24.   C
25.   C    BLOCK 2.  CONSTRUCT THE CONDUCTANCE MATRIX.
26.   C
27.   30    DO 100 K=1,NELEM
28.        IF(NLOOP.GT.0)GO TO 32
29.   C    NODE NUMBERS OF ELEMENT K.
30.        READ 31,NODE(K,1),NODE(K,2),NODE(K,3)
31.        PRINT 31,NODE(K,1),NODE(K,2),NODE(K,3)        (Continued)
```

Figure 6.11 (*Continued*)

```
 32.   31     FORMAT(3I10)
 33.   32     I=NODE(K,1)
 34.          J=NODE(K,2)
 35.          M=NODE(K,3)
 36.          A=0.5*((X(I)*Y(J)-X(J)*Y(I))+(X(M)*Y(I)-X(I)*Y(M))
 37.         1      +(X(J)*Y(M)-X(M)*Y(J)))
 38.   C    NX AND NY ARE SPATIAL DERIVATIVES OF INTERPOLATION FUNCTIONS.
 39.          NX(1)=0.5*(Y(J)-Y(M))/A
 40.          NX(2)=0.5*(Y(M)-Y(I))/A
 41.          NX(3)=0.5*(Y(I)-Y(J))/A
 42.          NY(1)=0.5*(X(M)-X(J))/A
 43.          NY(2)=0.5*(X(I)-X(M))/A
 44.          NY(3)=0.5*(X(J)-X(I))/A
 45.   C    SEEPAGE FACE IS IN ELEMENT 13.    NODES ARE 9,14,13.
 46.          IF(K.NE.13)GO TO 35
 47.          QX=NX(1)*H(I)+NX(2)*H(J)+NX(3)*H(M)
 48.          F(13)=QX*(Y(13)-Y(14))/2.
 49.          F(14)=F(13)
 50.   35     DO 40 KK=1,3
 51.          L=NODE(K,KK)
 52.          G(L,I)=G(L,I)+A*(NX(1)*NX(KK)+NY(1)*NY(KK))
 53.          G(L,J)=G(L,J)+A*(NX(2)*NX(KK)+NY(2)*NY(KK))
 54.          G(L,M)=G(L,M)+A*(NX(3)*NX(KK)+NY(3)*NY(KK))
 55.   40     CONTINUE
 56.  100     CONTINUE
 57.   C
 58.   C    BLOCK 3.    SOLVE SYSTEM OF EQUATIONS BY ITERATION.
 59.   C
 60.  200     AMAX=0.
 61.          DO 400 L=1,NNODE
```

```
C     EXCLUDE FIXED BOUNDARY HEADS FROM ITERATION.
      IF(KODE(L).EQ.1)GO TO 400
      OLDVAL=H(L)
      SUM=0.
      DO 300 JJ=1,NNODE
      IF(JJ.EQ.L)GO TO 300
      SUM=SUM+G(L,JJ)*H(JJ)
300   CONTINUE
      H(L)=(-SUM+F(L))/G(L,L)
      ERR=ABS(OLDVAL-H(L))
      IF(ERR.GT.AMAX)AMAX=ERR
400   CONTINUE
      IF(AMAX.GT.0.01)GO TO 200
C
C     BLOCK 4.  COMPARE TOP BOUNDARY HEADS WITH ASSUMED ELEVATIONS.
C
      AMAX=0.
      NLOOP=NLOOP+1
      DO 450 L=5,13,4
      OLDVAL=Y(L)
      Y(L)=H(L)
      ERR=ABS(OLDVAL-Y(L))
      IF(ERR.GT.AMAX)AMAX=ERR
450   CONTINUE
      IF(AMAX.GT.0.01)GO TO 30
      PRINT 605,NLOOP
605   FORMAT(1H1,'NUMBER OF ITERATIONS NLOOP = ',I5)
      PRINT 606
606   FORMAT(////,1X,'NODE NUMBER',6X,'HEAD')
      DO 610 L=1,NNODE
      PRINT 611,L,H(L)
611   FORMAT(I7,5X,F10.2)
610   CONTINUE
      STOP
      END
```

[143]

ECHO OF INPUT CARDS

```
      16            18
     .00          4.00         1        4.00
     .00          2.00         1        4.00
     .00          1.00         1        4.00
     .00           .00         1        4.00
    2.00          3.70         0         .00
    2.00          2.00         0         .00
    2.00          1.00         0         .00
    2.00           .00         0         .00
    4.00          3.36         0         .00
    4.00          2.00         0         .00
    4.00          1.00         0         .00
    4.00           .00         0         .00
    6.00          3.00         0         .00
    6.00          2.00         1        3.00
    6.00          1.00         1        3.00
    6.00           .00         1        3.00
       1             6         5
       1             2         6
       2             7         6
       2             3         7
       3             8         7
       3             4         8
       5            10         9
       5             6        10
       6            11        10
       6             7        11
       7            12        11
       7             8        12
       9            14        13
       9            10        14
      10            15        14
      10            11        15
      11            16        15
      11            12        16
```

Figure 6.12
Echo of input cards for free surface problem.

```
NUMBER OF ITERATIONS NLOOP =        4


    NODE NUMBER        HEAD
         1             4.00
         2             4.00
         3             4.00
         4             4.00
         5             3.72
         6             3.69
         7             3.68
         8             3.68
         9             3.40
        10             3.37
        11             3.36
        12             3.35
        13             3.08
        14             3.00
        15             3.00
        16             3.00
```

Figure 6.13
Final heads at all nodes for dam problem. The free surface is defined
by nodes 5, 9, and 13.

6.9 POISSON'S EQUATION

The results already obtained for Laplace's equation make it a straightforward
matter to extend the finite element technique to Poisson's equation. The Lth
equation of the system of algebraic equations is represented by the weighted
residual

$$\iint_D \left(\frac{\partial^2 \hat{h}}{\partial x^2} + \frac{\partial^2 \hat{h}}{\partial y^2} + \frac{R}{T} \right) N_L(x, y)\, dx\, dy = 0 \qquad (6.29)$$

where R is the recharge, and T is the transmissivity. We assume them to be
constant over the problem domain although they can have separate values for
each element.

We use the triangular elements and basis functions defined in Section 6.3. When compared with Equation 6.2, Equation 6.29 contains the additional term $(R/T)N_L$ in the integrand. If Equation 6.7 is applied for the second derivatives of $\hat{h}$, then

$$\iint_D \left(\frac{\partial \hat{h}}{\partial x} \frac{\partial N_L}{\partial x} + \frac{\partial \hat{h}}{\partial y} \frac{\partial N_L}{\partial y} \right) dx\,dy = \iint_D \frac{R}{T} N_L(x, y)\,dx\,dy$$

$$+ \int_\Gamma \left(\frac{\partial \hat{h}}{\partial x} n_x + \frac{\partial \hat{h}}{\partial y} n_y \right) N_L\,d\sigma \quad (6.30)$$

The integration of the recharge term on the right-hand side of Equation 6.30 is done over the problem domain by summing the integrations over elements.

$$\iint_D \frac{R}{T} N_L(x, y)\,dx\,dy = \sum_e \left(\iint_e \frac{R}{T} N_L(x, y)\,dx\,dy \right) \quad (6.31)$$

The recharge term leads to an additional column matrix, which we will call $\{B\}$; that is, Equation 6.16 becomes

$$[G]\{h\} = \{B\} + \{f\} \quad (6.32)$$

For an element e defined by nodes i, j, and m, $N_L = N_L^e$ for $L = i, j$ and m, and $N_L = 0$ for all other L. Therefore, the integration over element e on the right-hand side of Equation 6.31 contributes to three rows of $\{B\}$. These contributions can be represented by an element matrix

$$\{B^e\} = \begin{pmatrix} B_i^e \\ B_j^e \\ B_m^e \end{pmatrix} \quad (6.33)$$

where

$$B_i^e = \iint_e \frac{R}{T} N_i^e(x, y)\,dx\,dy = \frac{R}{T} \frac{A^e}{3} \quad (6.34a)$$

$$B_j^e = \iint_e \frac{R}{T} N_j^e(x, y)\,dx\,dy = \frac{R}{T} \frac{A^e}{3} \quad (6.34b)$$

$$B_m^e = \iint_e \frac{R}{T} N_m^e(x, y)\,dx\,dy = \frac{R}{T} \frac{A^e}{3} \quad (6.34c)$$

The area of element e is designated A^e. The evaluation of the integrals in Equation 6.34 follows from the property of N_L^e that its value at the centroid of a triangular element is one-third. The contribution of recharge within an element to each node is one-third of its integrated value. This intuitive result is analogous to Equation 6.23 in which one-half of the contribution of a boundary flux is distributed to each node bounding the side.

The global value B_L is the sum of the contributions B_L^e from each element.

$$B_L = \sum_e B_L^e \tag{6.35}$$

Computer Program

We create a new array $B(L)$ for the right-hand side of Equation 6.32. We obtain the contributions to $B(L)$ by summing the element contributions within the block of statements which constructs the element conductance and recharge matrices for $L = I, J$, and M. Therefore, after line 46 of Figure 6.6, we insert the statement

$$B(L) = B(L) + (R/T) * (A/3) \tag{6.36}$$

In the iterative solution for $H(L)$ in Block 3 of Figure 6.6, we must account for the nonzero right-hand side of Equation 6.32 and replace line 64 by

$$H(L) = (-SUM + B(L))/G(L, L) \tag{6.37}$$

Of course, the $B(L)$ array needs to be cleared and dimensioned elsewhere in the program. The program in Figure 6.6 can be modified further to incorporate variable T (heterogeneity) and variable recharge.

Notes and Additional Reading

1. An introduction to the finite element method using the calculus of variations is given by Remson et al. (1971, pp. 289–341). An intermediate-level discussion of finite element methods applied to both groundwater and surface water systems is contained in Cheng (1978), while an advanced-level text which treats both groundwater and surface water systems is Pinder and Gray (1977). Also, the finite element text by Brebbia and Connor (1977) deals with fluid flow. A general and readable treatment of finite element techniques is contained in Segerlind (1976) and Huebner (1975).

   Pinder and Gray (1977) contains a one-chapter review of differential equations and matrix algebra. The use of the Galerkin method is discussed on pp. 214–217 of Cheng (1978), and Pinder and Gray (1977) deal with the method of weighted residuals and the Galerkin method on pp. 54–57.

2. Applications of the Galerkin method may be found in many studies, including Pinder and Frind (1972) and Hodge and Freeze (1977).

3. The finite element technique is particularly well suited to analyzing flow through cross sections or profiles of the subsurface in which the position of the water table (the upper boundary) is unknown. This problem is analyzed by Neuman and Witherspoon (1970). Finnemore and Perry (1968) use a finite difference relaxation technique to solve the seepage problem through a dam of rectangular cross section.

4. The finite element development for Laplace's equation is easily generalized for the steady-state flow equation, Equation 2.17, in which the aquifer is allowed to be heterogeneous and anisotropic. Let K_x^e and K_y^e represent the principal hydraulic conductivities within element e. Then, a typical element conductance matrix term which would replace Equation 6.20a is

$$G_{L,i}^e = A^e \left(K_x^e \frac{\partial N_i^e}{\partial x} \frac{\partial N_L^e}{\partial x} + K_y^e \frac{\partial N_i^e}{\partial y} \frac{\partial N_L^e}{\partial y} \right) \qquad (6.38)$$

   Aquifer properties can vary from element to element, but in Equation 6.38 the heterogeneity is approximated as being constant within an element. An alternative formulation (Pinder et al., 1973) is to specify K_x and K_y at the nodes and to use the basis functions to interpolate the nodal values within an element.

5. Accurate keypunching and proofreading of input data are important practical considerations in handling large finite element meshes.

6. The same set of algebraic equations results from both the finite element and finite difference methods for certain cases in which the mesh of nodal points is regular. Such comparisons have been made by Zienkiewicz and Cheung (1965), Pinder and Gray (1976), and Wang and Anderson (1977).

7. The finite element method has its origin in problems of structural mechanics. Some hints in setting up the analogy are given in Appendix D (Table D.2).

Problems

6.1 Use Equation 6.11a to find the numerical values of

$$N_i^e(x_i, y_i), \ N_i^e(x_j, y_j), \text{ and } N_i^e(x_m, y_m).$$

6.2 Write algebraic expressions for the derivatives

$$\partial N_i^e/\partial x, \ \partial N_j^e/\partial x, \ \partial N_m^e/\partial x, \ \partial N_i^e/\partial y, \ \partial N_j^e/\partial y, \text{ and } \partial N_m^e/\partial y.$$

Compare the expressions with lines 33 to 38 of the computer program in Figure 6.6.

6.3 Construct the equation $L = 6$ for the finite element mesh shown in Figure 6.1b. The elements 1, 2, 3, 8, 9, and 10 all contain node 6 as a corner, and hence these six elements contribute to the summation in Equation 6.21 for each i that is a node in those elements.

6.4 Consider the finite element mesh of two elements and four nodes shown in Figure 6.14. Suppose the two element matrices are as follows:

$$[G^{e=1}] = \begin{array}{c} i=1 \\ j=2 \\ m=3 \end{array} \begin{pmatrix} i=1 & j=2 & m=3 \\ 3 & -4 & 7 \\ -4 & 1 & 2 \\ 7 & 2 & 1 \end{pmatrix} \qquad [G^{e=2}] = \begin{array}{c} i=3 \\ j=4 \\ m=1 \end{array} \begin{pmatrix} i=3 & j=4 & m=1 \\ 1 & 8 & 2 \\ 8 & 2 & -3 \\ 2 & -3 & 5 \end{pmatrix}$$

What is the global matrix $[G]$?

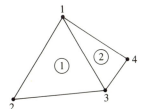

Figure 6.14
Finite element mesh for Problem 6.4.

6.5 Solve the well drawdown problem for the unknown heads h_6, h_7, h_{10}, and h_{11} using the Picasso-style finite element mesh (Figure 6.9). Compare the conductance matrix and answers with those in Figure 6.8.

6.6 Use the two-dimensional solution (Figure 6.13) to draw equipotential and flow lines through the dam cross section. How good are the Dupuit assumptions in this problem?

6.7 Find the position of the water table and the height of the seepage face for a dam with a width of 60 m. Compare the answers with those of the 6-m wide dam.

6.8 Consider the two equations

$$4h_1 + 3h_2 = 0$$
$$3h_1 - 7h_2 = 0$$

where h_1 is specified to be 5 and the first equation is really extraneous. Solve for h_2 by the method of changing the first equation to $4 \times 10^{10} h_1 + 3h_2 = 20 \times 10^{10}$.

6.9 Modify the program in Figure 6.6 to solve Poisson's equation. Incorporate the possibilities of heterogeneity (variable T) and variable recharge R.

6.10 Approximate a circular island by the finite element mesh shown in Figure 6.15. Use

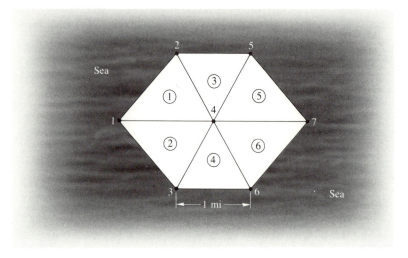

Figure 6.15
Finite element mesh for recharge on a circular island.

either a computer program such as the one from Problem 6.9, or write the single equation to calculate the potential at the center of the island. Take $R = 0.0031$ ft day^{-1} and $T = 10,000$ ft^2 day^{-1}. Compare the result with the analytical solution for a circular island

$$h(r) = \frac{R}{4T}(a^2 - r^2)$$

where $a = 1$ mi is the radius of the island and $r = 0$ is the location of the center. The variable r is the distance from the center of the island.

CHAPTER 7

Finite Elements: Transient Flow

7.1 INTRODUCTION

The application of the finite element method to the transient flow equation is discussed in this chapter. As in the case of steady-state flow, the spatial problem domain is subdivided into elements which are defined by nodes along their edges. However, in the transient flow case, the nodal values of head $h_L(t)$ are functions of time. The Galerkin method applied to the transient flow equation leads to a system of differential equations containing the first-order time derivatives $\partial h_L/\partial t$. The system of first-order differential equations can be written in matrix form. The stepping through time of the solution for the nodal heads is done by a finite difference approximation of the time derivative. What is usually called the finite element method of solving the transient flow equation is a hybrid of finite element and finite difference concepts.

In this chapter, we also introduce the interpolation functions for four-node rectangular elements. The rectangular element is very restrictive in accommodating irregular boundaries. However, the generalization to a quadrilateral as an isoparametric mapping of the rectangle is described in Appendix C. Isoparametric quadrilateral elements are quite useful in accommodating irregular boundaries.

7.2   GALERKIN'S METHOD

We apply Galerkin's method to the transient flow equation

$$\frac{\partial^2 h}{\partial x^2} + \frac{\partial^2 h}{\partial y^2} = \frac{S}{T}\frac{\partial h}{\partial t} \tag{7.1}$$

where S is the storage coefficient and T is the transmissivity. The procedural steps are identical to the ones in Chapter 6 for steady-state flow.

Trial Solution

We begin by assuming a trial solution $\hat{h}(x, y, t)$ of the form

$$\hat{h}(x, y, t) = \sum_{L=1}^{NNODE} h_L(t)N_L(x, y) \tag{7.2}$$

where $h_L(t)$ is the head at node L, $N_L(x, y)$ is the basis function associated with node L, and $NNODE$ is the total number of nodes in the problem domain. This form of the trial solution shows the separation between the space and time variables. At any instant of time, the basis functions interpolate the nodal heads over the problem domain.

Weighted Residual

Galerkin's method requires that, when the trial solution is substituted into the differential equation, the residual, when weighted by each of the basis functions, be zero.

$$\iint_D \left(\frac{\partial^2 \hat{h}}{\partial x^2} + \frac{\partial^2 \hat{h}}{\partial y^2} - \frac{S}{T} \frac{\partial \hat{h}}{\partial t} \right) N_L(x, y) \, dx \, dy = 0 \qquad (7.3)$$

where $L = 1, 2, \ldots, NNODE$; D signifies that the integration is done over the entire problem domain. Equation 7.3 generates one equation for each node.

Integration by Parts

The first spatial derivatives of the trial solution are discontinuous over element boundaries. Integration by parts reduces the order of derivatives by one. The integration by parts of the second spatial derivatives in Equation 7.3 is exactly as in Chapter 6 (Equation 6.7). Integration by parts is not required for the term containing $\partial \hat{h}/\partial t$ in the integrand of Equation 7.3. Therefore, Equation 7.3 can be written in the form

$$\iint_D \left(\frac{\partial \hat{h}}{\partial x} \frac{\partial N_L}{\partial x} + \frac{\partial \hat{h}}{\partial y} \frac{\partial N_L}{\partial y} \right) dx \, dy + \iint_D \frac{S}{T} \frac{\partial \hat{h}}{\partial t} N_L \, dx \, dy$$

$$= \int_\Gamma \left(\frac{\partial \hat{h}}{\partial x} n_x + \frac{\partial \hat{h}}{\partial y} n_y \right) N_L \, d\sigma \qquad (7.4)$$

where $L = 1, 2, \ldots, NNODE$, Γ is the boundary of the problem domain, n_x and n_y are the components of a unit vector normal to the boundary, and σ is an integration variable representing distance along the boundary in a counterclockwise sense. The boundary integral is a weighted average of the flux normal to the boundary.

7.3 RECTANGULAR ELEMENT

We now describe the implementation of Galerkin's method by defining the basis functions for rectangular elements. Each rectangle is defined by its four corner nodes. These are the points at which the heads $h_L(t)$ are unknowns. The same set of nodes can define both triangular and rectangular elements (Figure 7.1).

The archetypal rectangular element e is specified by the corner nodes i, j, m, and n labeled in counterclockwise order beginning with the lower left corner (Figure 7.2a). The archetypal rectangle is oriented parallel to the coordinate axes with its center at the origin. The element basis functions $N_L^e(x, y)$ define, in terms of the nodal heads $h_L(t)$, the trial solution $\hat{h}^e(x, y, t)$ over the element.

$$\hat{h}^e(x, y, t) = N_i^e(x, y)h_i(t) + N_j^e(x, y)h_j(t) + N_m^e(x, y)h_m(t) + N_n^e(x, y)h_n(t) \quad (7.5)$$

Figure 7.1

Two finite element meshes of the region-near-a-well. The same nodal points are used in both cases.

(a) Triangular elements.
(b) Rectangular elements.

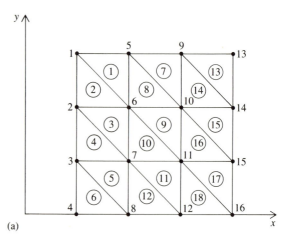

(a)

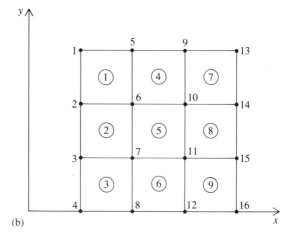

(b)

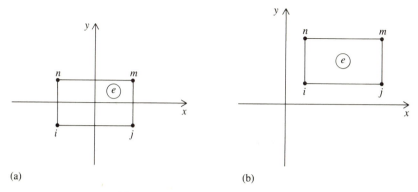

Figure 7.2
The archetypal rectangular element e.
(a) In local coordinates.
(b) In global coordinates.

where

$$N_i^e(x, y) = \frac{1}{4}\left(1 - \frac{x}{a}\right)\left(1 - \frac{y}{b}\right) \tag{7.6a}$$

$$N_j^e(x, y) = \frac{1}{4}\left(1 + \frac{x}{a}\right)\left(1 - \frac{y}{b}\right) \tag{7.6b}$$

$$N_m^e(x, y) = \frac{1}{4}\left(1 + \frac{x}{a}\right)\left(1 + \frac{y}{b}\right) \tag{7.6c}$$

$$N_n^e(x, y) = \frac{1}{4}\left(1 - \frac{x}{a}\right)\left(1 + \frac{y}{b}\right) \tag{7.6d}$$

and where

$$2a = x_j - x_i = x_m - x_n \tag{7.7a}$$

$$2b = y_n - y_i = y_m - y_j \tag{7.7b}$$

The four element basis functions for the archetypal rectangle are called bi-linear interpolations. Each element basis function has the property that $N_L^e(x, y)$ is 1 at node L, 0 at the other three nodes, and varies linearly with distance along any side. The signs within the parentheses of each equation for $N_L^e(x, y)$ are easily remembered because they correspond to the signs of x and y for the quadrant in which node L is located.

The element basis functions also define, in terms of the time derivatives $\partial h_L/\partial t$

at the nodes, the spatial variation of $\partial \hat{h}^e / \partial t$ over the element:

$$\frac{\partial \hat{h}^e}{\partial t} = N_i^e(x, y) \frac{\partial h_i}{\partial t} + N_j^e(x, y) \frac{\partial h_j}{\partial t} + N_m^e(x, y) \frac{\partial h_m}{\partial t} + N_n^e(x, y) \frac{\partial h_n}{\partial t} \quad (7.8)$$

The same interpolations are used for the time derivative of the trial solution as for the trial solution itself.

Local and Global Coordinates

The archetypal rectangle is centered at the origin of the coordinate system. The same definition of element basis functions can be applied to a rectangle not centered at the origin but whose sides are still parallel to the coordinate axes (Figure 7.2b). A simple translation to the origin of the coordinate system does not affect the value of an integration over the rectangle, provided the limits of integration are between $-a$ and $+a$ and between $-b$ and $+b$. The side lengths of $2a$ and $2b$ are given by Equation 7.7, assuming the rectangle is parallel to the coordinate axes. Before translation, the rectangle is said to be in global coordinates. After translation, the rectangle is said to be in local coordinates.

Patch of Elements

The collection of elements which contain a specific node L forms a patch around node L. Within the patch, the nodal basis function $N_L(x, y)$ is defined piecewise by the element basis functions $N_L^e(x, y)$ over each element belonging to the patch. Outside the patch, $N_L(x, y) = 0$. Thus, each basis function $N_L(x, y)$ is pyramidal in shape. Its peak value of one is located directly over node L. Its edges slope to zero at all the other nodes. Note that the patches and basis functions will be different for triangular and rectangular elements in the meshes of Figure 7.1.

7.4   ASSEMBLY OF MATRIX DIFFERENTIAL EQUATION

Equation 7.4 is the weighted residual condition for the Lth basis function. There is one such equation for each node L in the problem domain. We show in this section how the system of equations for $L = 1, 2, \ldots, NNODE$ can be written in matrix form as

$$[G]\{h\} + [P]\left\{\frac{\partial h}{\partial t}\right\} = \{f\} \qquad (7.9)$$

The three terms of the matrix equation can be compared with those of Equation 7.4 to obtain a sense of what each represents. The $[G]$ matrix is the conductance matrix which was developed for the steady-state case in Chapter 6. The $\{h\}$ matrix is a column matrix of nodal heads $h_L(t)$ at time t. The $[P]$ matrix is a square matrix which accounts for the storage term in the transient flow equation. The $\{\partial h/\partial t\}$ matrix is a column matrix of the time derivatives $\partial h_L/\partial t$. The $\{f\}$ matrix is a column matrix which represents the boundary conditions.

The integrals in Equation 7.4 can be evaluated element by element with the results summed over the problem domain

$$\sum_e \left\{ \iint_e \left(\frac{\partial \hat{h}^e}{\partial x}\frac{\partial N_L}{\partial x} + \frac{\partial \hat{h}^e}{\partial y}\frac{\partial N_L}{\partial y}\right) dx\,dy \right\} + \sum_e \left\{ \iint_e \frac{S}{T}\frac{\partial \hat{h}^e}{\partial t} N_L \, dx\,dy \right\}$$

$$= \int_\Gamma \left(\frac{\partial \hat{h}}{\partial x} n_x + \frac{\partial \hat{h}}{\partial y} n_y\right) N_L \, d\sigma \qquad (7.10)$$

where $L = 1, 2, \ldots, NNODE.$

Element Matrices

The subdivision of the problem domain into elements means that we evaluate the integrals in the left-hand side of Equation 7.10 one element at a time. A nodal basis function N_L is nonzero only over the patch of elements about node L. Therefore, the double integral over an element e is nonzero only if the element is in the patch about node L, in which case N_L is defined within the element to be the element basis function N_L^e. In other words, an element e made up of nodes i, j, m, and n will contribute only to those four equations when $L = i, j, m$, or n. These four node numbers make up the four row and four column labels of the element matrix.

If the element trial solution, Equation 7.5, is substituted into the first term of Equation 7.10, and if like coefficients of $h_i(t)$, $h_j(t)$, $h_m(t)$, and $h_n(t)$ are gathered, then the matrix entries for the Lth row of the element $[G]$ matrix are:

$$G^e_{L,i} = \int_{-a}^{a} \int_{-b}^{b} \left(\frac{\partial N^e_i}{\partial x} \frac{\partial N^e_L}{\partial x} + \frac{\partial N^e_i}{\partial y} \frac{\partial N^e_L}{\partial y} \right) dx\, dy \qquad (7.11a)$$

$$G^e_{L,j} = \int_{-a}^{a} \int_{-b}^{b} \left(\frac{\partial N^e_j}{\partial x} \frac{\partial N^e_L}{\partial x} + \frac{\partial N^e_j}{\partial y} \frac{\partial N^e_L}{\partial y} \right) dx\, dy \qquad (7.11b)$$

$$G^e_{L,m} = \int_{-a}^{a} \int_{-b}^{b} \left(\frac{\partial N^e_m}{\partial x} \frac{\partial N^e_L}{\partial x} + \frac{\partial N^e_m}{\partial y} \frac{\partial N^e_L}{\partial y} \right) dx\, dy \qquad (7.11c)$$

$$G^e_{L,n} = \int_{-a}^{a} \int_{-b}^{b} \left(\frac{\partial N^e_n}{\partial x} \frac{\partial N^e_L}{\partial x} + \frac{\partial N^e_n}{\partial y} \frac{\partial N^e_L}{\partial y} \right) dx\, dy \qquad (7.11d)$$

where $L = i, j, m,$ or n.

The element $[G]$ matrix is four by four because we are using a four-node element, and each node interacts with all the others as well as itself. In Chapter 6, the element $[G]$ matrix was three by three because we used a three-node element.

If the time derivative of the element trial solution, Equation 7.8, is substituted into the second term of Equation 7.10, and if like coefficients of $\partial h_i/\partial t$, $\partial h_j/\partial t$, $\partial h_m/\partial t$, and $\partial h_n/\partial t$ are gathered, then the matrix entries for the Lth row of the element $[P]$ matrix are:

$$P^e_{L,i} = \frac{S}{T} \int_{-a}^{a} \int_{-b}^{b} N^e_i N^e_L\, dx\, dy \qquad (7.12a)$$

$$P^e_{L,j} = \frac{S}{T} \int_{-a}^{a} \int_{-b}^{b} N^e_j N^e_L\, dx\, dy \qquad (7.12b)$$

$$P^e_{L,m} = \frac{S}{T} \int_{-a}^{a} \int_{-b}^{b} N^e_m N^e_L\, dx\, dy \qquad (7.12c)$$

$$P^e_{L,n} = \frac{S}{T} \int_{-a}^{a} \int_{-b}^{b} N^e_n N^e_L\, dx\, dy \qquad (7.12d)$$

where $L = i, j, m,$ or n.

Gaussian Quadrature

The integrations in Equations 7.11 and 7.12 could be done analytically as in Section 6.4. The integrands are polynomials in x and y, and the integrations are straightforward, although cumbersome. Numerical integration by Gaussian quadrature is easy to code in a computer program and, in this case, it leads to an exact numerical value for the integrations.

We first consider Gaussian quadrature in one dimension. Let $g(\xi)$ be a function defined for $-1 \leq \xi \leq 1$. The Gaussian quadrature procedure equates the definite integral to a weighted sum over a finite number of points

$$\int_{-1}^{1} g(\xi) \, d\xi = \sum_{i=1}^{n} W_i g(\xi_i) \tag{7.13}$$

where W_i is the required weighting factor at the Gauss point ξ_i. If $g(\xi)$ is a polynomial with terms up to ξ^2, Equation 7.13 is an exact equality by taking $n = 2$, $\xi_1 = -1/\sqrt{3} = -0.57735$, $\xi_2 = 1/\sqrt{3} = 0.57735$, and $W_1 = W_2 = 1$.

The double-integral generalization for a second-degree polynomial $g(\xi, \eta)$ is

$$\int_{-1}^{1} \int_{-1}^{1} g(\xi, \eta) \, d\xi \, d\eta = g(\xi_1, \eta_1) + g(\xi_2, \eta_1) + g(\xi_1, \eta_2) + g(\xi_2, \eta_2) \tag{7.14}$$

where the four Gauss points are defined by $\xi_1 = -1/\sqrt{3}$, $\xi_2 = 1/\sqrt{3}$, $\eta_1 = -1/\sqrt{3}$, and $\eta_2 = 1/\sqrt{3}$. All the weighting coefficients are equal to one.

The integrals in Equations 7.11 and 7.12 can be put in the form of Equation 7.14 through the following change of variables.

$$\xi = \frac{x}{a} \qquad \text{and} \qquad \eta = \frac{y}{b} \tag{7.15}$$

Then, $dx = a \, d\xi$ and $dy = b \, d\eta$. The limits of integration become -1 to 1 and the interpolation functions, Equation 7.6, in (ξ, η) coordinates are

$$\tilde{N}_i^e(\xi, \eta) = \tfrac{1}{4}(1 - \xi)(1 - \eta) \tag{7.16a}$$

$$\tilde{N}_j^e(\xi, \eta) = \tfrac{1}{4}(1 + \xi)(1 - \eta) \tag{7.16b}$$

$$\tilde{N}_m^e(\xi, \eta) = \tfrac{1}{4}(1 + \xi)(1 + \eta) \tag{7.16c}$$

$$\tilde{N}_n^e(\xi, \eta) = \tfrac{1}{4}(1 - \xi)(1 + \eta) \tag{7.16d}$$

In summary, the integrals in Equations 7.11 and 7.12 transform according to

$$\int_{-a}^{a} \int_{-b}^{b} g(x, y) \, dx \, dy = ab \int_{-1}^{1} \int_{-1}^{1} \tilde{g}(\xi, \eta) \, d\xi \, d\eta \tag{7.17}$$

where $\tilde{g}(\xi, \eta)$ is $g(x, y)$ expressed in (ξ, η) coordinates.

Global Matrices

The assembly of the global matrices $[G]$ and $[P]$ in Equation 7.9 is now straightforward. The element matrices are formed sequentially, and the contributions to the global matrices are summed. The contributions to the global matrices are sorted out by the subscripts in Equations 7.11 and 7.12 as the elements are considered one at a time. The process is the same as that illustrated in Figure 6.4, but the element matrices are now four by four.

Boundary Conditions

The matrix equation is completed by considering the boundary conditions which are represented by the column matrix $\{f\}$ in Equation 7.9. The Lth row of $\{f\}$ is given by the boundary integral on the right-hand side of Equation 7.10. This integral is identical to that in Equation 6.13, and its evaluation was described in Section 6.5. The results can be briefly summarized as follows:

1. $f_L = 0$ for all interior nodes.

2. $f_L = 0$ for nodes on no flow boundaries.

3. f_L is irrelevant for nodes on specified head boundaries if the Lth equation is eliminated.

7.5 SOLVING THE MATRIX DIFFERENTIAL EQUATION

Equation 7.9 is a first-order matrix differential equation. To solve it, we make a finite difference approximation for the time derivative in matrix notation

$$\left\{\frac{\partial h}{\partial t}\right\} = \frac{1}{\Delta t}\left(\{h\}^{t+\Delta t} - \{h\}^t\right) \tag{7.18}$$

where the superscript represents the time level and Δt is the length of the time step. Keep in mind that $\left\{\dfrac{\partial h}{\partial t}\right\}$ is a column matrix whose individual entries are $\partial h_L/\partial t$, where $h_L = h_L(t)$ is the value of the head at node L at time t. Thus the time derivative approximation at a particular node L is

$$\frac{\partial h_L}{\partial t} = \frac{h_L^{t+\Delta t} - h_L^t}{\Delta t} \tag{7.19}$$

Having made the finite difference approximation for $\left\{\dfrac{\partial h}{\partial t}\right\}$, we must specify some time level in the interval between t and $t + \Delta t$ at which to evaluate the term $[G]\{h\}$. The choice of time at which to evaluate $[G]\{h\}$ is exactly equivalent to the choice of time at which to evaluate the finite difference approximation of the spatial derivatives $\partial^2 h/\partial x^2$ and $\partial^2 h/\partial y^2$ in Chapter 4. If $\{h\}$ is approximated at the new time $t + \Delta t$, then the solution of Equation 7.9 is said to be fully implicit and is given by

$$[G]\{h\}^{t+\Delta t} + \frac{1}{\Delta t}[P](\{h\}^{t+\Delta t} - \{h\}^t) = \{f\} \tag{7.20}$$

Equation 7.20 can be rearranged to have all the heads at the old time to be on the right-hand side and all the heads at the new time to be on the left-hand side.

$$\left([G] + \frac{1}{\Delta t}[P]\right)\{h\}^{t+\Delta t} = \frac{1}{\Delta t}[P]\{h\}^t + \{f\} \tag{7.21}$$

Let the column matrix $\{B\}$ be equal to the first term on the right-hand side of Equation 7.21. The Lth row of $\{B\}$ can be explicitly calculated because heads at time t are known.

$$B_L = \frac{1}{\Delta t}\sum_{JJ=1}^{NNODE} P_{L,JJ}h^t_{JJ} \tag{7.22}$$

where JJ is the summation index and $NNODE$ is the number of nodes. Thus after the global matrices $[G]$ and $[P]$ are assembled and the initial conditions are set, heads at the end of the first time step are computed by solving Equation 7.21. These values are then used to compute an updated $\{B\}$ matrix and the bootstrap procedure continues.

Note that $[G]$ and $[P]$ need to be assembled only once for the whole problem, but the system of linear equations represented by Equation 7.21 must be solved at each time step. Note also that Equation 7.21 has basically the same form as Equation 6.16.

We have derived the matrix equation for the fully implicit case. In general, the heads $\{h\}$ can be approximated anywhere between t and $t + \Delta t$.

$$\{h\} = (1 - \alpha)\{h\}^t + \alpha\{h\}^{t+\Delta t} \tag{7.23}$$

where $0 \le \alpha \le 1$. If $\alpha = 1$, the solution is fully implicit. If $\alpha = 0$, the solution is fully explicit. The matrix equation similar to Equation 7.21, when $\{h\}$ is defined by Equation 7.23 and $\alpha = \frac{1}{2}$, is the Crank–Nicolson approximation.

7.6 COMPUTER PROGRAM FOR RESERVOIR PROBLEM

The computer program shown in Figure 7.3 specifically solves the transient reservoir lowering problem shown in Figure 4.1. The program is based on the

Figure 7.3

Finite element computer program to solve the transient flow reservoir lowering problem.

```
1.    C    FINITE ELEMENT PROGRAM FOR TRANSIENT FLOW USING RECTANGULAR ELEMENTS
2.    C    RESERVOIR EXAMPLE OF CHAPTER 4
3.         REAL NS,NX,NY
4.         DIMENSION HOLD(25),HNEW(25),G(25,25),P(25,25),B(25),X(25),Y(25)
5.         S=0.002
6.         T=0.02
7.         DIMENSION XSI(4),ETA(4),NS(4),NX(4),NY(4),NODE(4)
8.         DATA XSI/-.57735,.57735,.57735,-.57735/
9.         DATA ETA/-.57735,-.57735,.57735,.57735/
10.   C
11.   C    BLOCK 1.  GENERATE NODAL COORDINATES & INITIAL & BOUNDARY CONDITIONS.
12.   C
13.        NNODE=22
14.        NELEM=10
15.        DO 10 L=1,NNODE,2
16.        X(L)=(L-1)*5
17.        X(L+1)=X(L)
18.        Y(L)=10
19.   10   Y(L+1)=0
20.        DO 15 L=1,NNODE
21.        HOLD(L)=16.
22.        HNEW(L)=16.
23.        DO 15 JJ=1,NNODE
24.        G(L,JJ)=0
25.   15   P(L,JJ)=0
26.        DO 20 L=21,22
27.        HOLD(L)=11.
28.   20   HNEW(L)=11.
29.   C
```

```
30.  C    BLOCK 2.   CONSTRUCT CONDUCTANCE AND STORAGE MATRICES.
31.  C
32.       DO 100 K=1,NELEM
33.  C    GENERATE NODE NUMBERS OF ELEMENT K.
34.       I=2*K
35.       J=I+2
36.       M=I+1
37.       N=I-1
38.       NODE(1)=I
39.       NODE(2)=J
40.       NODE(3)=M
41.       NODE(4)=N
42.       AA=ABS(X(J)-X(I))/2
43.       BB=ABS(Y(N)-Y(I))/2
44.       DO 40 KK=1,4
45.       L=NODE(KK)
46.  C    GAUSSIAN QUADRATURE
47.       DO 30 IQ=1,4
48.  C    NS ARE INTERPOLATION FUNCTIONS.   NX AND NY ARE SPATIAL DERIVATIVES OF
49.       NS(1)=.25*(1-XSI(IQ))*(1-ETA(IQ))
50.       NS(2)=.25*(1+XSI(IQ))*(1-ETA(IQ))
51.       NS(3)=.25*(1+XSI(IQ))*(1+ETA(IQ))
52.       NS(4)=.25*(1-XSI(IQ))*(1+ETA(IQ))
53.       NX(1)=-.25*(1-ETA(IQ))/AA
54.       NX(2)=.25*(1-ETA(IQ))/AA
55.       NX(3)=.25*(1+ETA(IQ))/AA
56.       NX(4)=-.25*(1+ETA(IQ))/AA
```

(Continued)

Figure 7.3 *(Continued)*

```
57.        NY(1)=-.25*(1-XSI(IQ))/BB
58.        NY(2)=-.25*(1+XSI(IQ))/BB
59.        NY(3)=.25*(1+XSI(IQ))/BB
60.        NY(4)=.25*(1-XSI(IQ))/BB
61.        G(L,I)=G(L,I)+(NX(KK)+NY(1)*NX(KK)+NY(1)*NY(KK))*AA*BB
62.        G(L,J)=G(L,J)+(NX(2)*NX(KK)+NY(2)*NY(KK))*AA*BB
63.        G(L,M)=G(L,M)+(NX(3)*NX(KK)+NY(3)*NY(KK))*AA*BB
64.        G(L,N)=G(L,N)+(NX(4)*NX(KK)+NY(4)*NY(KK))*AA*BB
65.        P(L,I)=P(L,I)+NS(1)*NS(KK)*AA*BB*S/T
66.        P(L,J)=P(L,J)+NS(2)*NS(KK)*AA*BB*S/T
67.        P(L,M)=P(L,M)+NS(3)*NS(KK)*AA*BB*S/T
68.        P(L,N)=P(L,N)+NS(4)*NS(KK)*AA*BB*S/T
69.   30   CONTINUE
70.   40   CONTINUE
71.  100   CONTINUE
72. C
73. C   BLOCK  3.   STEP THROUGH TIME.
74. C
75.        PRINT 120
76.  120   FORMAT(1H1,31X,'HEAD',36X,'TIME',//)
77.        DT=5.
78.        KOUNT=1
79.        KPRINT=2
80.        TIME=DT
81.        DO 500 NSTEP=1,100
82. C   CONSTRUCT B-MATRIX FOR EACH TIME STEP.
83.        DO 150 L=1,NNODE
84.        B(L)=0
85.        DO 150 JJ=1,NNODE
86.        B(L)=B(L)+P(L,JJ)*HOLD(JJ)/DT
87.  150   CONTINUE
```

```
C     SOLVE SYSTEM OF EQUATIONS BY ITERATION.
200   AMAX=0
      DO 400 L=1,NNODE
C     EXCLUDE FIXED BOUNDARY HEADS FROM ITERATION.
      IF((L.EQ.1).OR.(L.EQ.2).OR.(L.EQ.21).OR.(L.EQ.22))GO TO 400
      OLDVAL=HNEW(L)
      SUM=0
      DO 300 JJ=1,NNODE
      IF(JJ.EQ.L)GO TO 300
      SUM=SUM+(G(L,JJ)+P(L,JJ)/DT)*HNEW(JJ)
300   CONTINUE
      HNEW(L)=(-SUM+B(L))/(G(L,L)+P(L,L)/DT)
      ERR=ABS(OLDVAL-HNEW(L))
      IF(ERR.GT.AMAX)AMAX=ERR
400   CONTINUE
      IF(AMAX.GT.0.01)GO TO 200
C     PUT HNEW VALUES INTO HOLD ARRAY FOR NEXT TIME STEP.
      DO 450 L=1,NNODE
450   HOLD(L)=HNEW(L)
C     PRINT RESULTS FOR EVERY OTHER TIME STEP.
      IF(KOUNT.NE.KPRINT)GO TO 490
      PRINT 401,(HNEW(I),I=1,21,2),TIME
401   FORMAT(1X,11F6.2,1F10.2)
      KOUNT=0
490   TIME=TIME+DT
      KOUNT=KOUNT+1
500   CONTINUE
      STOP
      END
```

[165]

```
88.
89.
90.
91.
92.
93.
94.
95.
96.
97.
98.
99.
100.
101.
102.
103.
104.
105.
106.
107.
108.
109.
110.
111.
112.
113.
114.
115.
116.
```

equations for rectangular elements developed in this chapter. The program has the same basic structure as the finite element program for Laplace's equation (Figure 6.6). Brief descriptions of the major program blocks follow.

Nodal Coordinates, Initial Conditions, and Boundary Conditions (Lines 11 to 28)

The finite element mesh for the problem is shown in Figure 7.4. The transient flow problem is one dimensional. Since we use two-dimensional rectangular elements, we impose no-flow conditions along the top and bottom boundaries. A line of boundary nodes with the corresponding values of f_L set to zero automatically satisfies the no-flow condition. The top and bottom values of head in any vertical column will be equal to each other as a result of no-flow conditions.

The regular array of nodal coordinates is easily generated in the program in the loop $DO\ 10\ L = 1,NNODE,2$. The old and new time step arrays, $HOLD(L)$ and $HNEW(L)$, are initialized to a value of 16 and the right boundary nodes are set to a value of 11 for both the old and new times.

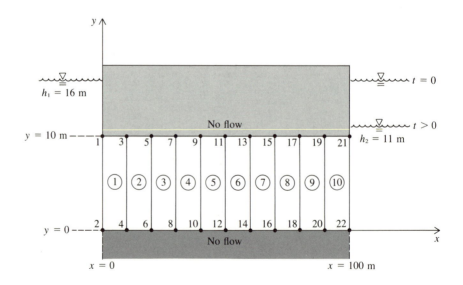

Figure 7.4
Finite element mesh for reservoir lowering problem.

Global Conductance and Storage Matrices (Lines 30 to 71)

The conductance matrix $[G]$ and storage matrix $[P]$ are constructed by summing the contributions from each element in sequence. The node numbers for the Kth element are generated by the program (lines 34 to 37), and the half-length AA and half-width BB of the element are calculated. The height in the y direction of each element is arbitrarily 10 m. In the loop DO 40 $KK = 1,4$, the program cycles through the cases $L = I$, $L = J$, $L = M$ and $L = N$. For each case, the element contributions, Equations 7.11 and 7.12, are computed and summed into the global $[G]$ and $[P]$ matrices (lines 61 to 68). The integrations are performed by Gaussian quadrature in the loop DO 30 $IQ = 1,4$ according to Equation 7.14. The change of variables to ξ and η requires that algebraic expressions be derived for the derivatives $\partial N_L^e / \partial x$ and $\partial N_L^e / \partial y$ in terms of ξ and η. The results are obtained by using the expressions for $\tilde{N}_L^e(\xi, \eta)$, Equation 7.16, and the chain rule of differentiation. The expressions are incorporated in lines 53 to 60. The interpolation functions NS and their space derivatives NX and NY, which go into the integrands according to the transformation, Equation 7.17, are evaluated at each of the four Gauss points and summed.

Time Stepping (Lines 73 to 114)

A profile of heads is computed for each time step in Block 3. First, the $\{B\}$ matrix for the fully implicit approximation, Equation 7.22, is obtained by matrix multiplication using the head values $HOLD(L)$ from the previous time step (lines 83 to 87). Next, the matrix equation, Equation 7.21, where $\{f\} = 0$, is solved by Gauss–Seidel iteration (lines 89 to 103). Lines 97 and 99 of Figure 7.3 replace lines 62 and 64 of Figure 6.6. Finally, the updated head values $HNEW(L)$ are placed into the array $HOLD(L)$ for the next time step. The answers are printed for every other time step (Figure 7.5) and can be compared with the finite difference answers (Figure 4.3).

Figure 7.5

Profile of head through time for the reservoir lowering problem as computed by the program in Figure 7.3.

HEAD											TIME
16.00	16.00	16.00	16.00	16.00	15.99	15.94	15.77	15.20	13.68	11.00	10.00
16.00	16.00	16.00	15.99	15.97	15.91	15.74	15.32	14.45	12.97	11.00	20.00
16.00	16.00	15.99	15.97	15.91	15.77	15.48	14.91	13.97	12.62	11.00	30.00
16.00	15.99	15.97	15.92	15.82	15.61	15.22	14.58	13.64	12.41	11.00	40.00
16.00	15.98	15.94	15.86	15.71	15.44	14.99	14.32	13.40	12.26	11.00	50.00
16.00	15.96	15.90	15.78	15.59	15.27	14.78	14.10	13.21	12.15	11.00	60.00
16.00	15.94	15.85	15.70	15.47	15.11	14.60	13.92	13.06	12.07	11.00	70.00
16.00	15.91	15.79	15.62	15.35	14.97	14.44	13.76	12.94	12.00	11.00	80.00
16.00	15.88	15.74	15.53	15.24	14.84	14.30	13.63	12.84	11.94	11.00	90.00
16.00	15.86	15.68	15.45	15.13	14.72	14.18	13.52	12.75	11.90	11.00	100.00
16.00	15.83	15.63	15.37	15.04	14.60	14.06	13.42	12.67	11.86	11.00	110.00
16.00	15.80	15.58	15.30	14.95	14.50	13.96	13.33	12.61	11.82	11.00	120.00
16.00	15.78	15.53	15.23	14.86	14.42	13.88	13.25	12.55	11.79	11.00	130.00
16.00	15.76	15.49	15.17	14.79	14.34	13.80	13.18	12.50	11.76	11.00	140.00
16.00	15.74	15.45	15.12	14.73	14.26	13.73	13.12	12.45	11.74	11.00	150.00
16.00	15.72	15.41	15.06	14.66	14.20	13.66	13.07	12.41	11.72	11.00	160.00
16.00	15.70	15.38	15.02	14.61	14.14	13.61	13.02	12.37	11.70	11.00	170.00
16.00	15.68	15.34	14.97	14.56	14.08	13.55	12.97	12.34	11.68	11.00	180.00
16.00	15.67	15.32	14.93	14.51	14.03	13.51	12.93	12.31	11.66	11.00	190.00
16.00	15.65	15.29	14.90	14.46	13.99	13.46	12.89	12.28	11.65	11.00	200.00
16.00	15.64	15.26	14.86	14.42	13.94	13.42	12.86	12.26	11.64	11.00	210.00
16.00	15.63	15.24	14.83	14.39	13.91	13.38	12.83	12.24	11.62	11.00	220.00
16.00	15.62	15.22	14.80	14.35	13.87	13.35	12.80	12.22	11.61	11.00	230.00

240.00	11.00	11.60	12.20	12.77	13.32	13.84	14.32	14.78	15.20	15.61	16.00
250.00	11.00	11.59	12.18	12.75	13.29	13.81	14.30	14.75	15.18	15.60	16.00
260.00	11.00	11.59	12.16	12.73	13.27	13.78	14.27	14.73	15.17	15.59	16.00
270.00	11.00	11.58	12.15	12.71	13.25	13.76	14.25	14.71	15.16	15.58	16.00
280.00	11.00	11.57	12.14	12.69	13.23	13.74	14.23	14.70	15.15	15.58	16.00
290.00	11.00	11.57	12.13	12.68	13.21	13.73	14.22	14.69	15.14	15.57	16.00
300.00	11.00	11.56	12.12	12.67	13.20	13.71	14.20	14.67	15.13	15.57	16.00
310.00	11.00	11.56	12.11	12.66	13.19	13.70	14.19	14.66	15.12	15.56	16.00
320.00	11.00	11.55	12.10	12.65	13.17	13.68	14.18	14.65	15.11	15.56	16.00
330.00	11.00	11.55	12.10	12.64	13.16	13.67	14.16	14.64	15.10	15.55	16.00
340.00	11.00	11.55	12.09	12.63	13.15	13.66	14.15	14.63	15.10	15.55	16.00
350.00	11.00	11.54	12.08	12.62	13.14	13.65	14.14	14.62	15.09	15.55	16.00
360.00	11.00	11.54	12.08	12.61	13.13	13.64	14.13	14.61	15.08	15.54	16.00
370.00	11.00	11.54	12.07	12.60	13.12	13.63	14.12	14.61	15.08	15.54	16.00
380.00	11.00	11.53	12.07	12.59	13.11	13.62	14.12	14.60	15.07	15.54	16.00
390.00	11.00	11.53	12.06	12.59	13.10	13.61	14.11	14.59	15.07	15.54	16.00
400.00	11.00	11.53	12.06	12.58	13.10	13.60	14.10	14.59	15.06	15.53	16.00
410.00	11.00	11.53	12.05	12.58	13.09	13.60	14.09	14.58	15.06	15.53	16.00
420.00	11.00	11.52	12.05	12.57	13.08	13.59	14.09	14.57	15.05	15.53	16.00
430.00	11.00	11.52	12.05	12.57	13.08	13.58	14.08	14.57	15.05	15.53	16.00
440.00	11.00	11.52	12.04	12.56	13.07	13.58	14.08	14.56	15.05	15.53	16.00
450.00	11.00	11.52	12.04	12.56	13.07	13.57	14.07	14.56	15.04	15.52	16.00
460.00	11.00	11.52	12.04	12.55	13.06	13.57	14.07	14.56	15.04	15.52	16.00
470.00	11.00	11.52	12.04	12.55	13.06	13.56	14.06	14.55	15.04	15.52	16.00
480.00	11.00	11.52	12.03	12.55	13.06	13.56	14.06	14.55	15.04	15.52	16.00
490.00	11.00	11.52	12.03	12.54	13.05	13.55	14.05	14.55	15.03	15.52	16.00
500.00	11.00	11.52	12.03	12.54	13.05	13.55	14.05	14.54	15.03	15.52	16.00

[169]

Notes and Additional Reading

1. In this chapter, only the spatial domain was divided into finite elements. The time dependence appeared as a first-order matrix differential equation with respect to time. This differential equation was numerically solved by finite differences. The time variable can also be included in the finite element formulation (Gray and Pinder, 1974).

2. A finite element model of transient flow is given with a good explanation of the technique by Durbin (1978). The application of the method to a field situation is described.

3. Nonsteady-state flow with a free surface is discussed by Neuman and Witherspoon (1971) and by Neuman (1973).

Problems

7.1   Use Equations 7.15 and 7.16 and the chain rule of differentiation to find expressions in (ξ, η) coordinates for $\partial N_L^e / \partial x$ and $\partial N_L^e / \partial y$, where $L = i, j, m$, and n. Compare the expressions with lines 53 to 60 of the computer program in Figure 7.3.

7.2   Derive the result that if $g(\xi) = a_0 + a_1 \xi + a_2 \xi^2$, then

$$\int_{-1}^{1} g(\xi) \, d\xi = g\left(\frac{-1}{\sqrt{3}}\right) + g\left(\frac{1}{\sqrt{3}}\right)$$

7.3   Assemble the Lth row of the conductance matrix $[G]$ for $L = 6$ for rectangular elements in Figure 7.1b. Compare the result with that for triangular elements from Chapter 6.

7.4   Explicitly obtain algebraic expressions for $G_{i,i}^e$ and $P_{i,i}^e$; that is, analytically perform the required integrations in Equations 7.11a and 7.12a.

7.5   For the reservoir problem (Figure 7.4), analytically integrate the element matrix terms $G_{2,2}^e$ and $P_{2,2}^e$ for those elements in the patch about node 2. Then explicitly compute the global matrix terms $G_{2,2}$ and $P_{2,2}$. Print out the global matrices in the computer program, Figure 7.3, and compare your answers.

7.6   Consider the following FORTRAN statements:

```
DO 25 KKK=1,4
II=NODE(KKK)
G(L,II)=G(L,II)+(NX(II)*NX(KK)+NY(II)*NY(KK))*AA*BB
P(L,II)=P(L,II)+NS(II)*NS(KK)*AA*BB*S/T
25    CONTINUE
```

For what lines in Figure 7.3 can these statements be substituted? If these statements are used, then we no longer need the $IJMN$ notation and lines 34 to 41 can be simplified to:

$$NODE(1) = 2*K$$
$$NODE(2) = NODE(1) + 2$$
$$NODE(3) = NODE(1) + 1$$
$$NODE(4) = NODE(1) - 1$$

7.7 Derive the matrix equation similar to Equation 7.21 for the Crank–Nicolson approximation. Modify and run the computer program for the reservoir problem (Figure 7.3) to use the Crank–Nicolson approximation.

7.8 Solve, using rectangular elements, the steady-state region-near-a-well problem in which the boundary heads were specified (Section 2.2). Compare the answers with the ones found in Chapter 6 using triangular elements.

7.9 Analytically evaluate the expression

$$B_L^e = \frac{R}{T} \int_{-a}^{a} \int_{-b}^{b} N_L^e(x, y)\, dx\, dy$$

where

$$N_L^e(x, y) = \frac{1}{4}\left(1 + \frac{x}{a}\right)\left(1 - \frac{y}{b}\right)$$

The result is the element contribution of one rectangle to the Lth row of $\{B\}$ in the matrix equation $[G]\{h\} = \{B\}$ for the finite element formulation of Poisson's equation.

7.10 Solve for the free surface of a 6-m wide dam (Figure 6.10) by using nine isoparametric quadrilateral elements (Appendix C). Incorporate the modification for isoparametric quadrilateral elements (Figure C.2) into the program in Figure 6.11. Remember to $DIMENSION$ and make $REAL$ the arrays $NXSI$ and $NETA$. Also, because Figure 6.11 was for triangular elements, modifications must be made to read in the fourth node number of each element, and to handle the flow boundary condition on the seepage face.

CHAPTER 8

Advective-Dispersive Transport

8.1 INTRODUCTION

The transport at the same velocity as groundwater of dissolved solids or of heat is called advective transport. We use the term advection in preference to the term convection because, strictly speaking, convection refers to fluid motion caused by temperature differences. In the case of heat transport in an aquifer, both advection and convection can occur simultaneously. The movement of contaminants in groundwater is a particularly active area of research. Models have been developed to study saltwater intrusion as well as leachate migration from waste disposal sites. See Anderson (1979) for a literature review of models that have been used to simulate solute transport in groundwater systems. The movement of waste heat from power plants and the movement of heat in groundwater systems in geothermal areas, as well as in areas where the use of groundwater heat pumps might be feasible, have also been studied by means of models. See, respectively, Andrews and Anderson (1979), Mercer et al. (1975), and Andrews (1978) for examples.

Movement of contaminants in groundwater occurs not only by advection but also by dispersion. Dispersion refers to mixing and spreading caused in part by molecular diffusion and in part by variations in velocity within the porous medium. For many field problems, dispersion caused by molecular diffusion and by flow around grains in the porous medium is negligible in comparison with dispersion caused by large-scale heterogeneities within the aquifer. In the presence of large scale heterogeneities, dispersion occurs as contaminants move selectively around less permeable units. If a tracer slug is injected instantaneously into a uniform flow field, it disperses in the direction of flow (longitudinally) and transverse to the direction of flow (laterally). There is greater dispersion in the direction of flow than transverse to the direction of flow. In an idealized picture, the initial point becomes an elliptically shaped cloud whose concentration drops off from the center following a Gaussian distribution (Figure 8.1). The center of the cloud represents the initial point carried solely by advective transport. The cloud itself is the result of dispersion.

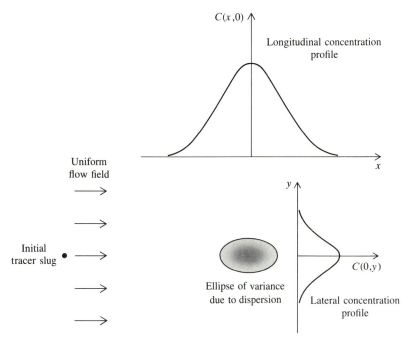

Figure 8.1
Dispersion of an instantaneous point source in a uniform flow field. The longitudinal and lateral distributions of concentration in the ellipse are shown by the superimposed graphs.

8.2 DISPERSION

Dispersion of a contaminant in groundwater is due mainly to heterogeneity of the medium. Dispersion is a result of the existence of a statistical distribution of flow paths and of flow velocities around local heterogeneities (Figure 8.2).

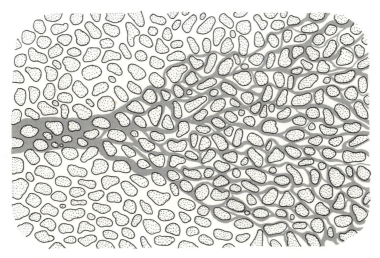

Figure 8.2
Statistical distribution of flow paths around local heterogeneities leads to dispersion. The process is shown here at a microscopic scale where pore space surrounds gravel-sized grains. (From R. A. Freeze and J. A. Cherry, *Groundwater*,©1979, p. 384. Reprinted by permission of Prentice-Hall, Inc., Englewood Cliffs, N. J.)

Laboratory experiments have demonstrated that microscopic dispersion caused by flow around grains of a porous medium is on the order of centimeters, whereas field studies suggest that dispersion caused by macroscopic heterogeneities (for example, clay lenses within a sandy aquifer) is on the order of meters.

The standard technique for measuring in situ hydraulic conductivity involves the analysis of aquifer or pumping test drawdown data. This technique yields a spatially averaged value for the area affected by the test. Although spatially averaged values are usually satisfactory for flow problems, the distribution of heterogeneities is important in controlling the movement of contaminants. For example, experiments have demonstrated that tracers injected into a porous medium flow more readily through stringers of more permeable material, and the distribution of these stringers governs the magnitude of the dispersion. The presence of lenses of less permeable material embedded in more permeable material also results in greater dispersion as the tracer is deflected around the less permeable units. Thus spatially averaged values of hydraulic conductivity are not adequate for describing dispersion.

Dispersion results in the smearing of an initially sharp concentration front as individual contaminant particles travel along different paths and at different velocities. Dispersion is illustrated by a classical experiment in which a tracer is continuously introduced at the upgradient end of a laboratory sand column through which water is flowing under steady-state conditions. The tracer concentration is expressed as a relative concentration C/C_0, where C is the concentration of the tracer at the outlet and C_0 is the initial concentration of the inflow. The graph of concentration versus time at a particular spatial point is known as the breakthrough curve (Figure 8.3). If the tracer moves through the column with no dispersion, the breakthrough curve will be a step function.

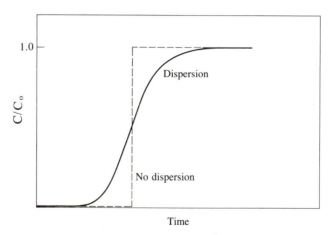

Figure 8.3
Breakthrough curves when dispersion is present and when it is not present.

Dispersive Flux

Consider a flow field in which the average linear or pore velocity has components v_x and v_y. The tracer is carried along by the flow field and is dispersed. The microscopic velocity at any point in the aquifer may vary from the average linear velocity. Let v_x^* and v_y^* be the microscopic velocity components relative to the average velocity components v_x and v_y; that is, v_x^* and v_y^* represent dispersive tracer movement. Their definition incorporates the idea that point-to-point statistical fluctuations of velocity exist relative to the average.

A mass flux is the amount of mass crossing a unit cross-sectional area per unit time. The dispersive mass flux is the mass flux which results from the relative velocity components v_x^* and v_y^*. In an infinitesimal volume about a point in the aquifer, the groundwater specific discharge through a face perpendicular to the x direction is nv_x^*, where n is the porosity. If C is the solute concentration as a mass per unit volume of water, then the dispersive mass flux in the x direction is

$$f_x^* = nCv_x^* \tag{8.1}$$

Similarly, the dispersive flux in the y direction is

$$f_y^* = nCv_y^* \tag{8.2}$$

Dispersion Coefficient (Uniform Flow Field)

We define the longitudinal and transverse components of the dispersion coefficient for a uniform flow field in the x direction, where $v_x = \bar{v}_x$ and $v_y = 0$. The results can then be generalized. The longitudinal component D_L of the dispersion coefficient is defined by analogy to Fick's law of diffusion. Specifically, the dispersive flux is assumed to be proportional to the concentration gradient in the flow direction times porosity.

$$nCv_x^* = -nD_L \frac{\partial C}{\partial x} \tag{8.3}$$

Dividing through by n gives

$$Cv_x^* = -D_L \frac{\partial C}{\partial x} \tag{8.4}$$

Likewise, the transverse component D_T of the dispersion coefficient is defined such that the dispersive flux perpendicular to the flow direction is proportional to the concentration gradient in the y direction times porosity.

$$nCv_y^* = -nD_T \frac{\partial C}{\partial y} \tag{8.5}$$

Dividing through by n gives

$$Cv_y^* = -D_T \frac{\partial C}{\partial y} \tag{8.6}$$

Equations 8.4 and 8.6 are similar to Fick's law of diffusion, because the distribution of velocities about the mean at each point in the groundwater flow field is assumed to be analogous to the distribution of molecular velocities about the mean in a diffusing gas.

Dispersivity

Experiments have demonstrated that, in an isotropic medium, the longitudinal and transverse components of dispersion in Equations 8.4 and 8.6 are linearly dependent on the average speed of groundwater flow. For a uniform flow field with an average linear velocity equal to $\bar{v}_x$,

$$D_L = a_L \bar{v}_x \tag{8.7}$$

and

$$D_T = a_T \bar{v}_x \tag{8.8}$$

where the parameters a_L and a_T are the longitudinal and transverse dispersivities, respectively. The dispersivities are intended to be an intrinsic physical property of the porous medium and have units of length. One would expect the dispersivities to vary spatially with changes in the lithology of the porous medium. However, dispersivities generally have been assumed to be constants in models.

Recently, the practice of using constant dispersivities has come under scrutiny. Smith and Schwartz (1980) suggest that it is essential to account for spatial variability in dispersivity to arrive at a correct assessment of contaminant movement. They also maintain that if the velocity field could be defined in detail, it would be possible to use laboratory (microscopic) values for dispersivity rather than the high values generally used in field studies. Many investigators have pointed out that dispersivity appears to vary according to the scale of the problem. Dispersivity values are determined in the field by monitoring the movement of a tracer, and the value calculated depends on the distance traveled by the tracer. At larger scales, a greater number of heterogeneities are encountered, and a higher dispersivity is needed to account for the observed dispersion. Gelhar et al. (1979) use stochastic analysis to demonstrate that, in theory, dispersivity should approach a constant value for long travel times and large distances from the contaminant source.

Dispersion Coefficient (Nonuniform Flow Field)

The dispersion coefficient has directional properties which we have described with respect to a uniform groundwater flow field in the x direction. Let us now consider a nonuniform flow field where the groundwater velocity vector components are v_x and v_y. The velocity is a function of position. We define two separate coordinate systems. We select one convenient global coordinate system for the entire problem domain. Then, at each point in the aquifer, we define a local coordinate system (x', y') such that the x' axis coincides with the direction of the velocity $\mathbf{v}$ (Figure 8.4). The counterclockwise rotation angle θ from global to local coordinates is defined at each point by $\tan \theta = v_y/v_x$.

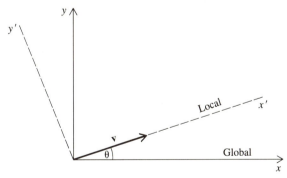

Figure 8.4
Definition of local coordinate system in which velocity vector parallels x' axis. The axes of the local coordinate system are in the principal directions of the dispersion coefficient tensor. The longitudinal and lateral components can be related back to the global coordinates of the problem by Equation 8.17.

In the local coordinate system, the velocity vector parallels the x' axis. Therefore, the previous development for the longitudinal and transverse dispersion coefficients in a uniform flow field can be used; that is, in the local coordinate system,

$$D_L = a_L|\mathbf{v}| \tag{8.9}$$

and

$$D_T = a_T|\mathbf{v}| \tag{8.10}$$

where

$$\mathbf{v} = \sqrt{v_x^2 + v_y^2} \tag{8.11}$$

The dispersion coefficient is the proportionality constant which relates the mass flux vector to the gradient of concentration. The dispersion coefficient, like hydraulic conductivity, is a second rank tensor (see Appendix A). The principal directions are in the direction of flow and perpendicular to the direction of flow. In a nonuniform flow field, the principal directions vary from point to point in the problem domain. In the local coordinate system, which coincides with the principal directions, the dispersion tensor is diagonal; that is,

$$[D'] = \begin{pmatrix} D_L & 0 \\ 0 & D_T \end{pmatrix} \tag{8.12}$$

It is necessary to transform the dispersion coefficient tensor from its local coordinate system into the global coordinate system of the problem domain. The global coordinate system will not, in general coincide with the principal axes at each point. For nonprincipal axes, the generalization of the law relating mass flux to concentration gradient, Equations 8.4 and 8.6, is

$$Cv_x^* = -D_{11}\frac{\partial C}{\partial x} - D_{12}\frac{\partial C}{\partial y} \tag{8.13}$$

$$Cv_y^* = -D_{21}\frac{\partial C}{\partial x} - D_{22}\frac{\partial C}{\partial y} \tag{8.14}$$

where

$$[D] = \begin{pmatrix} D_{11} & D_{12} \\ D_{21} & D_{22} \end{pmatrix} \tag{8.15}$$

Rotating the dispersion coefficient tensor into global coordinates is the same as rotating the hydraulic conductivity tensor (Equation A.10)

$$[D] = [R]^{-1}[D'][R]$$ (8.16)

where $[R]$ is the rotation matrix for the counterclockwise rotation angle from global to local coordinates. The components of $[D]$ resulting from the matrix multiplication in Equation 8.16 are

$$D_{11} = D_L \cos^2 \theta + D_T \sin^2 \theta$$ (8.17a)

$$D_{22} = D_L \sin^2 \theta + D_T \cos^2 \theta$$ (8.17b)

$$D_{12} = D_{21} = (D_L - D_T) \sin \theta \cos \theta$$ (8.17c)

8.3 SOLUTE TRANSPORT EQUATION

The assumed law that the dispersive mass flux is proportional to the concentration gradient (Equations 8.4 and 8.6 for a uniform flow field) is known as the Fickian model. This law plays the same role in obtaining a governing equation for advective-dispersive mass transport as does Darcy's law in obtaining the governing equation for groundwater flow. The other law used in the derivation is the continuity or conservation of mass principle.

We derive the time dependent governing equation for the case of steady-state uniform flow in the x direction. The generalization to a two-dimensional nonuniform flow field will then be apparent. The total mass flux in the x direction is the sum of the advective and dispersive fluxes.

$$f_x = n(C\bar{v}_x + Cv_x^*)$$ (8.18)

The assumption of uniform flow in the x direction means that the total mass flux in the y direction is just the dispersive flux.

$$f_y = nCv_y^*$$ (8.19)

The advective term in Equation 8.18 is $nC\bar{v}_x$, and the dispersive terms in Equations 8.18 and 8.19 are nCv_x^* and nCv_y^*. Continuity or conservation requires that the divergence of the flux, that is, the net outward flow of mass per unit volume of the aquifer per unit time, be equal to the rate of decrease of solute concentration per unit volume of aquifer.

$$\frac{\partial f_x}{\partial x} + \frac{\partial f_y}{\partial y} = -n\frac{\partial C}{\partial t} \tag{8.20}$$

The time derivative of the solute concentration is multiplied by the porosity so that the right-hand side of Equation 8.20 represents the change in solute mass per unit aquifer volume rather than per unit water volume. Recall that the concentration is defined as solute mass per unit water volume. Combining Equations 8.4, 8.6, 8.18, 8.19, and 8.20 gives

$$\frac{\partial}{\partial x}\left(nD_L\frac{\partial C}{\partial x}\right) + \frac{\partial}{\partial y}\left(nD_T\frac{\partial C}{\partial y}\right) - \frac{\partial}{\partial x}(nC\bar{v}_x) = n\frac{\partial C}{\partial t} \tag{8.21}$$

The generalization to arbitrary two-dimensional flow requires the full tensor form of the dispersion coefficient and follows from Equations 8.13 and 8.14.

$$\frac{\partial}{\partial x}\left(nD_{11}\frac{\partial C}{\partial x} + nD_{12}\frac{\partial C}{\partial y}\right) + \frac{\partial}{\partial y}\left(nD_{21}\frac{\partial C}{\partial x} + nD_{22}\frac{\partial C}{\partial y}\right)$$

$$-\frac{\partial}{\partial x}(nCv_x) - \frac{\partial}{\partial y}(nCv_y) = n\frac{\partial C}{\partial t} \tag{8.22}$$

where at each point in the aquifer, D_{ij} values are given by Equation 8.17 as the dispersion coefficient rotated from local to global coordinates. Equations 8.21 and 8.22 carry the assumptions that sources or sinks are absent, and that no chemical reactions occur between solute and the porous medium.

   Many investigators have noted that the advection-dispersion governing equation just derived does not accurately simulate contaminant transport when the travel distance is short. Specifically, breakthrough curves measured in the laboratory do not fit the curves predicted from theory when this form of the equation is used. Gelhar et al. (1979) use stochastic analysis to derive a revised form of the advection-dispersion equation, which demonstrates that the Fickian model for dispersive transport is not adequate near the source of the contaminant. The revised form of the equation has extra terms which are important for short travel distances, but they become insignificant for long travel distances (large times). Non-Fickian transport is also discussed by Matheron and De Marsily (1980).

Sources, Sinks, and Chemical Reactions

Equation 8.22 can be readily modified to allow for sources and sinks and for certain kinds of chemical reactions. To include sources or sinks, the term $-C'W/b$ is added to the left-hand side of Equation 8.22, where C' is the concentration of solute in the source or sink fluid, W is the volume flow rate per unit aquifer area (positive for outflow and negative for inflow), and b is the thickness of the aquifer.

A generalized form of the chemical reaction term to be added to the left-hand side of Equation 8.22 is $n \sum_{k=1}^{s} R_k$, where R_k is the rate of production of the solute in reaction k of s different reactions (positive for addition of solute and negative for removal). An important example is the adsorption of solute on the solid surfaces within the porous medium. Adsorption and desorption act as sinks and sources. The process is usually rapid enough to be considered an ion exchange reaction in chemical equilibrium. The chemical reaction term for such a reaction is $-\rho_b(\partial \bar{C}/\partial t)$, where ρ_b is the bulk density of the dry material (mass of solid per unit volume of aquifer), and $\bar{C}$ is the relative concentration of the solute adsorbed (mass of solute per unit mass of bulk dry porous material). The adsorbed contaminant $\bar{C}$ is a function of the solute concentration C. Relationships between $\bar{C}$ and C represent the chemical equilibrium of the process and are known as adsorption isotherms because they are determined experimentally in the laboratory at a constant temperature. If $\bar{C} = K_d C$, where K_d is the distribution coefficient, then the chemical reaction term becomes $-\rho_b K_d(\partial C/\partial t)$. The adsorption term is thus written explicitly in terms of the solute concentration C. Anderson (1979) discusses the use of distribution coefficients as well as other expressions used to quantify ion exchange.

Solving the Governing Equation

There are a host of analytical solutions to the advection-dispersion equation under a variety of simplifying assumptions and boundary and initial conditions. Some of these solutions are reviewed by Anderson (1979). Because of the simplifying assumptions needed to obtain an analytical solution, most are not directly applicable to field situations.

Three numerical techniques have been used regularly to solve the advection-dispersion equation. Early solutions used finite difference approximations. However, these solutions sometimes contain numerical errors which invalidate the solution. Numerical errors can be minimized when the method of characteristics is used in conjunction with finite difference approximations. In the method of characteristics, the partial differential equation for advective-dispersive transport is first replaced by a set of ordinary differential equations, and then these are approximated using finite differences. More recently, interest has shifted to the use of finite element techniques to approximate the advection-dispersion equation. Finite element solutions are, in general, more immune to numerical errors than are finite difference solutions.

Figure 8.5 illustrates the two types of numerical errors which in some degree plague all numerical solutions of the advection-dispersion equation. For small

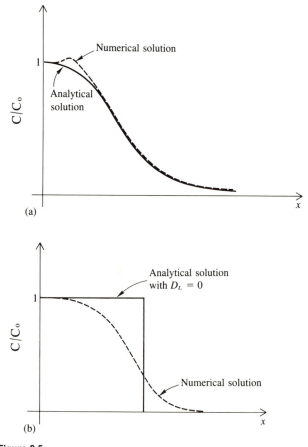

Figure 8.5
Numerical difficulties illustrated for a one-dimensional solution to the advective-dispersive transport equation. (a) Effect of overshoot. (b) Effect of numerical dispersion. (Adapted with permission from *Critical Reviews in Environmental Control* 9(2), 1979. Copyright The Chemical Rubber Co., CRC Press, Inc. After Fried, 1975.)

coefficients of dispersion, overshoot can be reduced by decreasing either the space interval or the time increment. Numerical dispersion can be minimized by adding a correction term to erase some of the numerical dispersion. Anderson (1979) discusses these numerical difficulties in more detail.

8.4 FINITE ELEMENT EXAMPLE: SOLUTE DISPERSION IN UNIFORM FLOW FIELD

The finite element computer program developed in Section 7.6 for the transient problem involving the lowering of water level in a reservoir is readily modified to solve a classical, one-dimensional solute dispersion problem. We consider a uniform groundwater flow field in the x direction, where $\bar{v}_x$ is the average linear velocity. For time t less than or equal to zero, the solute concentration throughout the aquifer is zero. For times greater than zero, the concentration at the left edge of the aquifer becomes a constant C_0. These initial and boundary conditions could simulate the sudden dumping of chloride or some other non-reactive (conservative) constituent into the reservoir lake after steady-state uniform flow has been achieved. Equation 8.21 is the governing equation to be used to predict the change in chloride concentrations with time at different distances x through the aquifer. Note that $D_T = 0$ in Equation 8.21 because the problem is one dimensional. In mathematical terms, the initial and boundary conditions of this problem are

Initial condition: $C(x, 0) = 0$ for all x

Boundary conditions: $C(0, t) = C_0$ for $t > 0$

$C(\infty, t) = 0$ for $t > 0$

The analytical solution has been derived by Ogata and Banks (1961).

$$C(x, t) = \frac{C_0}{2} \left\{ \exp\left(\frac{\bar{v}_x x}{D_L}\right) \operatorname{erfc}\left(\frac{x + \bar{v}_x t}{2\sqrt{D_L t}}\right) + \operatorname{erfc}\left(\frac{x - \bar{v}_x t}{2\sqrt{D_L t}}\right) \right\} \quad (8.23)$$

where $\bar{v}_x$ is the average linear velocity, D_L is the longitudinal component of the dispersion coefficient, and

$$\operatorname{erfc}(z) = \frac{2}{\sqrt{\pi}} \int_z^\infty e^{-u^2} \, du \quad (8.24)$$

The function erfc(z) is the complementary error function. The nature of the solution, Equation 8.23, is shown in Figure 8.6. For a given velocity—for example, 0.1 m day^{-1}—the breakthrough front is sharper for lower values of the dispersion coefficient, that is, for lower dispersivity.

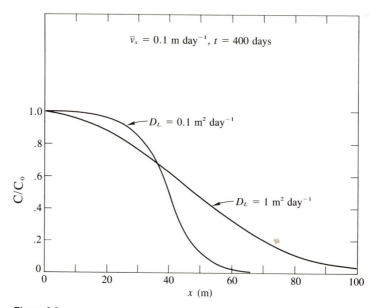

Figure 8.6
One-dimensional longitudinal dispersion showing effect of changing the value of the dispersion coefficient.

Finite Element Theory

The dispersion coefficients and porosity will be independent of position for a homogeneous medium in a uniform flow field. For this case, Equation 8.21 becomes

$$D_L \frac{\partial^2 C}{\partial x^2} + D_T \frac{\partial^2 C}{\partial y^2} - \bar{v}_x \frac{\partial C}{\partial x} = \frac{\partial C}{\partial t} \tag{8.25}$$

We apply the Galerkin method as in Chapter 7 for four-node rectangular elements. The residuals of the trial solution $\hat{C}$, weighted by the nodal basis functions $N_L(x, y)$, are set to zero

$$\iint_D \left(D_L \frac{\partial^2 \hat{C}}{\partial x^2} + D_T \frac{\partial^2 \hat{C}}{\partial y^2} - \bar{v}_x \frac{\partial \hat{C}}{\partial x} - \frac{\partial \hat{C}}{\partial t} \right) N_L(x, y) \, dx \, dy = 0 \qquad (8.26)$$

where $L = 1, 2, \ldots, NNODE$. The trial solution within an element $ijmn$ is an interpolation of the nodal values

$$\hat{C}^e(x, y, t) = N_i^e(x, y) C_i(t) + N_j^e(x, y) C_j(t) + N_m^e(x, y) C_m(t) + N_n^e(x, y) C_n(t) \qquad (8.27)$$

where the element nodal basis functions $N_L^e(x, y)$ are defined by Equation 7.6.

The second spatial derivative terms are integrated by parts and the integration over the problem domain is done element by element.

$$\sum_e \left\{ \iint_e \left(D_L \frac{\partial \hat{C}^e}{\partial x} \frac{\partial N_L}{\partial x} + D_T \frac{\partial \hat{C}^e}{\partial y} \frac{\partial N_L}{\partial y} + \bar{v}_x \frac{\partial \hat{C}^e}{\partial x} N_L + \frac{\partial \hat{C}^e}{\partial t} N_L \right) dx \, dy \right\}$$

$$= \int_\Gamma \left(D_L \frac{\partial \hat{C}}{\partial x} n_x + D_T \frac{\partial \hat{C}}{\partial y} n_y \right) N_L \, d\sigma \qquad (8.28)$$

where $L = 1, 2, \ldots, NNODE$.

The system of equations represented by Equation 8.28 can be written in matrix notation in the form

$$[G]\{C\} + [U]\{C\} + [P]\left\{ \frac{\partial C}{\partial t} \right\} = \{f\} \qquad (8.29)$$

where $\{C\}$ is the column matrix of nodal concentrations, and $\left\{ \frac{\partial C}{\partial t} \right\}$ is the column matrix of the time derivative of nodal concentrations. The square coefficient matrices $[G]$, $[U]$, and $[P]$ correspond to individual terms in the integral on the left-hand side of Equation 8.28. The column matrix $\{f\}$ corresponds to the boundary integral on the right-hand side of Equation 8.28.

Without the advective term, Equation 8.25 for mass transport is very similar in form to the transient flow equation, Equation 7.1. Therefore, the $[G]$ and $[P]$ matrices in Equation 8.29 have their nearly identical counterparts from those in Equations 7.11 and 7.12. For the mass transport equation under consideration in this chapter, typical element matrix entries are

$$G_{L,i}^e = \int_{-a}^{a} \int_{-b}^{b} \left(D_L \frac{\partial N_i^e}{\partial x} \frac{\partial N_L^e}{\partial x} + D_T \frac{\partial N_i^e}{\partial y} \frac{\partial N_L^e}{\partial y} \right) dx \, dy \qquad (8.30)$$

$$P_{L,i}^e = \int_{-a}^{a} \int_{-b}^{b} N_i^e N_L^e \, dx \, dy \qquad (8.31)$$

The matrix $[U]$ derives from the advective term $\bar{v}_x(\partial\hat{C}^e/\partial x)N_L$ in the integrand of Equation 8.28. Taking the x derivative of Equation 8.27 yields

$$\frac{\partial\hat{C}^e}{\partial x} = \frac{\partial N_i^e}{\partial x}C_i + \frac{\partial N_j^e}{\partial x}C_j + \frac{\partial N_m^e}{\partial x}C_m + \frac{\partial N_n^e}{\partial x}C_n \qquad (8.32)$$

Therefore, the coefficient which multiplies C_i is

$$U_{L,i}^e = \bar{v}_x \int_{-a}^{a}\int_{-b}^{b} \frac{\partial N_i^e}{\partial x} N_L^e \, dx \, dy \qquad (8.33a)$$

Similarly, the remaining element matrix terms are

$$U_{L,j}^e = \bar{v}_x \int_{-a}^{a}\int_{-b}^{b} \frac{\partial N_j^e}{\partial x} N_L^e \, dx \, dy \qquad (8.33b)$$

$$U_{L,m}^e = \bar{v}_x \int_{-a}^{a}\int_{-b}^{b} \frac{\partial N_m^e}{\partial x} N_L^e \, dx \, dy \qquad (8.33c)$$

$$U_{L,n}^e = \bar{v}_x \int_{-a}^{a}\int_{-b}^{b} \frac{\partial N_n^e}{\partial x} N_L^e \, dx \, dy \qquad (8.33d)$$

The evaluation of the integrals is done by Gaussian quadrature as described in Section 7.4. The assembly of the global matrices is performed by summation of the element terms. Note that the matrix equation, Equation 8.29, has the form

$$[A]\{C\} + [P]\left\{\frac{\partial C}{\partial t}\right\} = \{f\} \qquad (8.34)$$

where $[A] = [G] + [U]$. The entries of $[G]$ are symmetric in their subscripts. However, the entries to $[U]$ from the advective term, Equation 8.33, are not symmetric in their subscripts. The solution of Equation 8.34 through time is done by a finite difference approximation for $\left\{\dfrac{\partial C}{\partial t}\right\}$ in the same manner as described in Section 7.5.

Finite Element Computer Program

The finite element computer program to solve the solute transport problem for a uniform flow field requires only a few straightforward modifications of the computer program for transient flow (Figure 7.3). The rectangular finite element mesh of Figure 7.4 is used for the one-dimensional solute transport problem. The boundary conditions for $t > 0$ are chosen to be $C(0, t) = C_0 = 10$ and $C(100, t) = 0$. The right boundary condition at $x = 100$ m is the approximation to the boundary condition $C(\infty, t) = 0$. The finite element solution is limited to times less than the time needed for the solute front to travel about two-thirds of the 100-m width. The no-flow boundary conditions are kept as no-transport boundary conditions, thereby making the solute transport problem one-dimensional. The initial condition is $C(x, 0) = 0$ for all x.

The basic structure of the computer program shown in Figure 8.7 consists of three blocks. In Block 1, the nodal coordinates are generated and the initial and boundary conditions are set. In Block 2, the global coefficient matrices are generated by summing the contributions from each element. The integrations required by Equations 8.30, 8.31, and 8.33 are done by Gaussian quadrature. In Block 3, the solution is stepped through time. The matrix differential equation, Equation 8.34, is approximated by a fully implicit scheme. At each time step, the system of linear equations is solved by Gauss–Seidel iteration. The basic unknown of solute concentration is represented by the variables *COLD* and *CNEW* at the old and new time steps, respectively.

A few specific remarks about changes from the program in Chapter 7 follow:

Block 1. The flow velocity $\bar{v}_x = VX = 0.1$ m day^{-1}, the longitudinal dispersion coefficient $D_L = DX = 1$ m^2 day^{-1}, and the transverse dispersion coefficient $D_T = DY = 0.1$ m^2 day^{-1} are set in lines 12 to 14. The value of the transverse dispersion coefficient is immaterial for this one-dimensional problem, but it is inserted because the $[G]$ matrix in the program is for two-dimensional problems. The initial and boundary conditions are given in lines 22 to 24 and lines 29 to 31, respectively.

Block 2. Lines 64 to 71 are the only really significant changes to the old program. Lines 64 to 67 are just like lines 61 to 64 of Figure 7.3, but they incorporate components of the dispersion coefficient. Lines 68 to 71 represent element matrix values for the advective term.

Block 3. Line 83 changes the heading of the output to read *CONCEN-TRATION* instead of *HEAD*.

The output of the finite element program is shown in Figure 8.8. This numerical solution can be compared with the analytical solution which is shown in Figure 8.9. Note that the finite element solution predicts that the solute front travels somewhat faster than the analytical solution does.

```
1.      C     FINITE ELEMENT PROGRAM FOR SOLUTE TRANSPORT
2.      C     MODIFIED FROM RESERVOIR PROGRAM, FIG. 7.3
3.            REAL NS,NX,NY
4.            DIMENSION COLD(25),CNEW(25),G(25,25),P(25,25),U(25,25)
5.            DIMENSION B(25),X(25),Y(25)
6.            DIMENSION XSI(4),ETA(4),NS(4),NX(4),NY(4),NODE(4)
7.            DATA XSI/-.57735,.57735,.57735,-.57735/
8.            DATA ETA/-.57735,-.57735,.57735,.57735/
9.      C
10.     C     BLOCK 1.     GENERATE NODAL COORDINATES & INITIAL & BOUNDARY CONDITIONS.
11.     C
12.           VX=0.1
13.           DX=10.*VX
14.           DY=VX
15.           NNODE=22
16.           NELEM=10
17.           DO 10 L=1,NNODE,2
18.           X(L)=(L-1)*5
19.           X(L+1)=X(L)
20.           Y(L)=10
21.    10     Y(L+1)=0
22.           DO 15 L=1,NNODE
23.           COLD(L)=0.
24.           CNEW(L)=0.
25.           DO 15 JJ=1,NNODE
26.           G(L,JJ)=0
27.           U(L,JJ)=0
28.    15     P(L,JJ)=0
29.           DO 20 L=1,2
30.           COLD(L)=10.
31.    20     CNEW(L)=10.
32.     C
```

```
C     BLOCK 2.    CONSTRUCT GLOBAL COEFFICIENT MATRICES.
C
C     GENERATE NODE NUMBERS OF ELEMENT K.
      DO 100 K=1,NELEM
      I=2*K
      J=I+2
      M=I+1
      N=I-1
      NODE(1)=I
      NODE(2)=J
      NODE(3)=M
      NODE(4)=N
      AA=ABS(X(J)-X(I))/2
      BB=ABS(Y(N)-Y(I))/2
      DO 40 KK=1,4
      L=NODE(KK)
C     GAUSSIAN QUADRATURE
      DO 30 IQ=1,4
C     NS ARE INTERPOLATION FUNCTIONS.  NX AND NY ARE SPATIAL DERIVATIVES OF
      NS(1)=.25*(1-XSI(IQ))*(1-ETA(IQ))
      NS(2)=.25*(1+XSI(IQ))*(1-ETA(IQ))
      NS(3)=.25*(1+XSI(IQ))*(1+ETA(IQ))
      NS(4)=.25*(1-XSI(IQ))*(1+ETA(IQ))
      NX(1)=-.25*(1-ETA(IQ))/AA
      NX(2)=.25*(1-ETA(IQ))/AA
      NX(3)=.25*(1+ETA(IQ))/AA
      NX(4)=-.25*(1+ETA(IQ))/AA
      NY(1)=-.25*(1-XSI(IQ))/BB
      NY(2)=-.25*(1+XSI(IQ))/BB
```

(Continued)

[191]

Figure 8.7 *(Continued)*

```
62.        NY(3)=.25*(1+XSI(IQ))/BB
63.        NY(4)=.25*(1-XSI(IQ))/BB
64.        G(L,I)=G(L,I)+(DX*NX(1)*NX(KK)+DY*NY(1)*NY(KK))*AA*BB
65.        G(L,J)=G(L,J)+(DX*NX(2)*NX(KK)+DY*NY(2)*NY(KK))*AA*BB
66.        G(L,M)=G(L,M)+(DX*NX(3)*NX(KK)+DY*NY(3)*NY(KK))*AA*BB
67.        G(L,N)=G(L,N)+(DX*NX(4)*NX(KK)+DY*NY(4)*NY(KK))*AA*BB
68.        U(L,I)=U(L,I)+VX*NX(1)*NS(KK)*AA*BB
69.        U(L,J)=U(L,J)+VX*NX(2)*NS(KK)*AA*BB
70.        U(L,M)=U(L,M)+VX*NX(3)*NS(KK)*AA*BB
71.        U(L,N)=U(L,N)+VX*NX(4)*NS(KK)*AA*BB
72.        P(L,I)=P(L,I)+NS(1)*NS(KK)*AA*BB
73.        P(L,J)=P(L,J)+NS(2)*NS(KK)*AA*BB
74.        P(L,M)=P(L,M)+NS(3)*NS(KK)*AA*BB
75.        P(L,N)=P(L,N)+NS(4)*NS(KK)*AA*BB
76.   30   CONTINUE
77.   40   CONTINUE
78.  100   CONTINUE
79. C
80. C  BLOCK  3.  STEP THROUGH TIME.
81. C
82.        PRINT 120
83.  120   FORMAT(1H1,26X,'CONCENTRATION',32X,'TIME',//)
84.        DT=5.
85.        KOUNT=1
86.        KPRINT=2
87.        TIME=DT
88.        DO 500 NSTEP=1,100
89. C  CONSTRUCT B-MATRIX FOR EACH TIME STEP.
90.        DO 150 L=1,NNODE
91.        B(L)=0
92.        DO 150 JJ=1,NNODE
93.        B(L)=B(L)+P(L,JJ)*COLD(JJ)/DT
94.  150   CONTINUE
```

```
C     SOLVE SYSTEM OF EQUATIONS BY ITERATION.
200   AMAX=0
      DO 400 L=1,NNODE
C     EXCLUDE FIXED BOUNDARY CONCENTRATIONS FROM ITERATION.
      IF((L.EQ.1).OR.(L.EQ.2).OR.(L.EQ.21).OR.(L.EQ.22))GO TO 400
      OLDVAL=CNEW(L)
      SUM=0
      DO 300 JJ=1,NNODE
      IF(JJ.EQ.L)GO TO 300
      SUM=SUM+(G(L,JJ)+U(L,JJ)+P(L,JJ)/DT)*CNEW(JJ)
300   CONTINUE
      CNEW(L)=(-SUM+B(L))/(G(L,L)+U(L,L)+P(L,L)/DT)
      ERR=ABS(OLDVAL-CNEW(L))
      IF(ERR.GT.AMAX)AMAX=ERR
400   CONTINUE
      IF(AMAX.GT.0.01)GO TO 200
C     PUT CNEW VALUES INTO COLD ARRAY FOR NEXT TIME STEP.
      DO 450 L=1,NNODE
450   COLD(L)=CNEW(L)
C     PRINT RESULTS FOR EVERY OTHER TIME STEP.
      IF(KOUNT.NE.KPRINT)GO TO 490
      PRINT 401,(CNEW(I),I=1,21,2),TIME
401   FORMAT(1x,11F6.2,1F10.2)
      KOUNT=0
490   TIME=TIME+DT
      KOUNT=KOUNT+1
500   CONTINUE
      STOP
      END
```

95.
96.
97.
98.
99.
100.
101.
102.
103.
104.
105.
106.
107.
108.
109.
110.
111.
112.
113.
114.
115.
116.
117.
118.
119.
120.
121.
122.
123.

Figure 8.8
Profile of concentration through time for the solute transport problem as computed by the program in Figure 8.7.

CONCENTRATION										TIME
10.00	1.85	-.11	-.00	.00	-.00	.00	.00	.00	.00	10.00
10.00	3.17	.19	-.04	.00	.00	.00	.00	.00	.00	20.00
10.00	4.16	.66	-.03	-.01	.00	.00	.00	.00	.00	30.00
10.00	4.93	1.20	.07	-.02	.00	.00	.00	.00	.00	40.00
10.00	5.54	1.75	.23	-.01	-.00	.00	.00	.00	.00	50.00
10.00	6.04	2.29	.46	.02	-.01	.00	.00	.00	.00	60.00
10.00	6.46	2.80	.72	.08	-.01	-.00	.00	.00	.00	70.00
10.00	6.80	3.27	1.02	.17	.00	-.00	.00	.00	.00	80.00
10.00	7.10	3.71	1.33	.29	.03	-.00	.00	.00	.00	90.00
10.00	7.36	4.12	1.66	.44	.06	-.00	.00	.00	.00	100.00
10.00	7.59	4.50	1.98	.61	.12	.01	.00	.00	.00	110.00
10.00	7.79	4.85	2.31	.80	.19	.02	.00	.00	.00	120.00
10.00	7.96	5.17	2.63	1.01	.27	.04	.00	.00	.00	130.00
10.00	8.12	5.47	2.94	1.22	.37	.08	.01	.00	.00	140.00
10.00	8.26	5.74	3.25	1.45	.49	.12	.02	.00	.00	150.00
10.00	8.38	5.99	3.54	1.68	.62	.17	.03	.00	.00	160.00
10.00	8.50	6.23	3.82	1.92	.76	.23	.05	.01	.00	170.00
10.00	8.60	6.45	4.10	2.15	.91	.30	.07	.01	.00	180.00
10.00	8.69	6.65	4.36	2.39	1.08	.39	.11	.02	.01	190.00
10.00	8.78	6.84	4.61	2.63	1.24	.48	.14	.03	.01	200.00
10.00	8.85	7.02	4.85	2.86	1.42	.58	.19	.05	.01	210.00
10.00	8.93	7.18	5.08	3.10	1.60	.69	.24	.07	.01	220.00
10.00	8.99	7.33	5.30	3.33	1.79	.81	.30	.09	.02	230.00

240.00	.00	.03	.12	.37	.93	1.97	3.55	5.50	7.48	9.05	10.00
250.00	.00	.04	.15	.44	1.07	2.16	3.77	5.70	7.61	9.11	10.00
260.00	.00	.06	.19	.52	1.20	2.36	3.98	5.89	7.73	9.16	10.00
270.00	.00	.08	.24	.61	1.35	2.55	4.19	6.07	7.85	9.21	10.00
280.00	.00	.10	.29	.71	1.50	2.74	4.40	6.24	7.96	9.25	10.00
290.00	.00	.12	.34	.81	1.65	2.93	4.59	6.41	8.06	9.29	10.00
300.00	.00	.15	.40	.91	1.80	3.12	4.78	6.56	8.16	9.33	10.00
310.00	.00	.18	.47	1.02	1.96	3.31	4.97	6.71	8.25	9.37	10.00
320.00	.00	.22	.54	1.14	2.12	3.49	5.15	6.86	8.34	9.41	10.00
330.00	.00	.26	.62	1.26	2.28	3.68	5.33	6.99	8.42	9.44	10.00
340.00	.00	.31	.70	1.39	2.45	3.86	5.50	7.12	8.50	9.47	10.00
350.00	.00	.35	.78	1.51	2.61	4.04	5.66	7.25	8.58	9.50	10.00
360.00	.00	.41	.87	1.65	2.77	4.21	5.82	7.37	8.65	9.53	10.00
370.00	.00	.46	.97	1.78	2.94	4.38	5.97	7.48	8.71	9.55	10.00
380.00	.00	.52	1.07	1.92	3.10	4.55	6.12	7.59	8.77	9.57	10.00
390.00	.00	.58	1.17	2.05	3.26	4.72	6.26	7.69	8.83	9.60	10.00
400.00	.00	.65	1.28	2.19	3.42	4.88	6.40	7.79	8.89	9.62	10.00
410.00	.00	.72	1.39	2.34	3.58	5.04	6.53	7.88	8.94	9.64	10.00
420.00	.00	.79	1.50	2.48	3.74	5.19	6.66	7.97	8.99	9.65	10.00
430.00	.00	.87	1.61	2.62	3.90	5.34	6.78	8.06	9.04	9.67	10.00
440.00	.00	.94	1.73	2.77	4.05	5.48	6.90	8.14	9.08	9.69	10.00
450.00	.00	1.02	1.85	2.91	4.21	5.63	7.02	8.22	9.12	9.70	10.00
460.00	.00	1.11	1.97	3.05	4.36	5.76	7.13	8.29	9.16	9.72	10.00
470.00	.00	1.19	2.09	3.20	4.50	5.90	7.23	8.36	9.20	9.73	10.00
480.00	.00	1.28	2.22	3.34	4.65	6.03	7.33	8.43	9.24	9.74	10.00
490.00	.00	1.37	2.34	3.48	4.79	6.16	7.43	8.50	9.27	9.76	10.00
500.00	.00	1.45	2.47	3.62	4.93	6.28	7.53	8.56	9.30	9.77	10.00

Figure 8.9
Profile of concentration through time for the solute transport problem as computed from the analytical solution, Equation 8.23.

CONCENTRATION										TIME
10.00	.41	.00	.00	.00	.00	.00	.00	.00	.00	10.00
10.00	1.81	.04	.00	.00	.00	.00	.00	.00	.00	20.00
10.00	3.10	.25	.00	.00	.00	.00	.00	.00	.00	30.00
10.00	4.11	.64	.03	.00	.00	.00	.00	.00	.00	40.00
10.00	4.90	1.13	.11	.00	.00	.00	.00	.00	.00	50.00
10.00	5.53	1.66	.24	.02	.00	.00	.00	.00	.00	60.00
10.00	6.05	2.19	.44	.05	.00	.00	.00	.00	.00	70.00
10.00	6.47	2.71	.68	.10	.01	.00	.00	.00	.00	80.00
10.00	6.83	3.19	.95	.18	.02	.00	.00	.00	.00	90.00
10.00	7.14	3.65	1.26	.28	.04	.00	.00	.00	.00	100.00
10.00	7.40	4.07	1.57	.41	.07	.01	.00	.00	.00	110.00
10.00	7.63	4.46	1.90	.57	.12	.03	.00	.00	.00	120.00
10.00	7.83	4.82	2.23	.75	.18	.05	.00	.00	.00	130.00
10.00	8.00	5.16	2.55	.94	.25	.08	.00	.00	.00	140.00
10.00	8.16	5.46	2.87	1.15	.35	.11	.00	.00	.00	150.00
10.00	8.30	5.75	3.19	1.37	.45	.16	.01	.00	.00	160.00
10.00	8.42	6.01	3.49	1.61	.57	.21	.02	.00	.00	170.00
10.00	8.54	6.25	3.78	1.84	.71	.28	.04	.00	.00	180.00
10.00	8.64	6.47	4.06	2.08	.86	.36	.07	.01	.00	190.00
10.00	8.73	6.68	4.33	2.32	1.01	.44	.10	.01	.00	200.00
10.00	8.81	6.87	4.59	2.57	1.18	.54	.13	.02	.00	210.00
10.00	8.89	7.05	4.84	2.81	1.35	.64	.17	.03	.01	220.00
10.00	8.96	7.22	5.07	3.04	1.53	.64	.22	.04	.01	230.00

240.00	.00	.01	.05	.28	.76	1.72	3.28	5.30	7.37	9.03	10.00
250.00	.00	.02	.07	.34	.88	1.91	3.51	5.51	7.52	9.09	10.00
260.00	.01	.03	.09	.41	1.01	2.10	3.73	5.72	7.65	9.14	10.00
270.00	.01	.03	.17	.49	1.14	2.30	3.95	5.91	7.78	9.19	10.00
280.00	.01	.04	.22	.57	1.29	2.49	4.17	6.09	7.89	9.24	10.00
290.00	.02	.06	.26	.66	1.43	2.69	4.38	6.27	8.00	9.28	10.00
300.00	.02	.07	.31	.76	1.58	2.88	4.58	6.44	8.11	9.32	10.00
310.00	.03	.09	.37	.86	1.74	3.07	4.78	6.60	8.20	9.36	10.00
320.00	.04	.11	.43	.97	1.90	3.27	4.97	6.75	8.30	9.39	10.00
330.00	.05	.13	.50	1.08	2.06	3.46	5.15	6.89	8.38	9.43	10.00
340.00	.06	.16	.57	1.20	2.22	3.64	5.33	7.03	8.46	9.46	10.00
350.00	.07	.19	.65	1.33	2.39	3.83	5.50	7.16	8.54	9.49	10.00
360.00	.09	.22	.73	1.45	2.55	4.01	5.67	7.28	8.61	9.51	10.00
370.00	.10	.26	.82	1.58	2.72	4.19	5.83	7.40	8.68	9.54	10.00
380.00	.12	.30	.92	1.72	2.89	4.36	5.98	7.51	8.74	9.56	10.00
390.00	.14	.50	1.01	1.86	3.05	4.54	6.13	7.62	8.80	9.58	10.00
400.00	.17	.56	1.12	2.00	3.22	4.70	6.28	7.72	8.85	9.60	10.00
410.00	.20	.63	1.22	2.14	3.38	4.87	6.42	7.82	8.91	9.62	10.00
420.00	.23	.70	1.33	2.28	3.55	5.03	6.55	7.91	8.96	9.64	10.00
430.00	.26	.78	1.44	2.43	3.71	5.18	6.68	8.00	9.01	9.66	10.00
440.00	.30	.86	1.56	2.57	3.87	5.33	6.80	8.08	9.05	9.68	10.00
450.00	.33	.94	1.68	2.72	4.02	5.48	6.92	8.16	9.09	9.69	10.00
460.00	.38	1.03	1.80	2.86	4.18	5.63	7.03	8.24	9.14	9.71	10.00
470.00	.42	1.12	1.92	3.01	4.33	5.77	7.14	8.31	9.17	9.72	10.00
480.00	.47	1.22	2.05	3.16	4.48	5.90	7.25	8.38	9.21	9.73	10.00
490.00	.52	1.32	2.17	3.30	4.63	6.03	7.35	8.45	9.25	9.74	10.00
500.00	.57	1.42	2.30	3.45	4.78	6.16	7.45	8.51	9.28	9.76	10.00

Additional Reading

1. For a more rigorous derivation of the generalized advection-dispersion equation, see Bredehoeft and Pinder (1973), Konikow and Grove (1977), or Freeze and Cherry (1979, pp. 549–553). The mathematical nature of the dispersion coefficient and dispersivity is discussed in detail by Bear (1972).

2. The application of finite element methods to the advection-dispersion equation is discussed by Cheng (1978, pp. 209–210, 217–220, and 249–252), and by Pinder and Gray (1977, pp. 142–183). Also see Gray (1976) and Grove (1977).

3. Some of the problems involved in analyzing contaminant transport in groundwater are discussed by Cherry et al. (1975). Field techniques for determining dispersivity are described by Fried (1975), Webster et al. (1970), Ivanovich and Smith (1978), and Pickens et al. (1980) and are summarized by Anderson (1979), who also presents tables of dispersivity values that have been measured in the field or used in numerical models.

4. The use of the finite element method to solve the one-dimensional mass transport equation is discussed by Guymon (1970) and Desai (1979). De Marsily et al. (1977) discuss an application of a finite difference solution of a one-dimensional mass transport model to the migration of radioactive contaminants from a subsurface repository.

5. An application of a two-dimensional finite element profile model to a hypothetical problem is presented by Pickens and Lennox (1976). Pinder (1973) uses a two-dimensional areal finite element model with isoparametric quadrilaterals to simulate the migration of chromium from a waste disposal site in Long Island. Gureghian et al. (1980) use a two-dimensional finite element model to study movement of chloride from a landfill in Long Island. Segol et al. (1975) and Segol and Pinder (1976) use a two-dimensional finite element profile model to study saltwater encroachment along the coast of Florida.

6. Several investigators, including Pinder and Cooper (1970), Konikow and Bredehoeft (1974), Schwartz (1975, 1977), Konikow (1977), and Robson (1978), use the method of characteristics, together with finite differences, to solve the advection-dispersion equation in two dimensions and apply the model to a field problem. A well-documented version of a method of characteristics model is developed by Konikow and Bredehoeft (1978).

7. The problem of incorporating chemical reaction terms into advection-dispersion models is considered by Grove (1976), Back and Cherry (1976), Pickens and Lennox (1976), and Rubin and James (1973).

Problems

8.1 Modify the program in Figure 8.7 to solve the one-dimensional solute transport problem using the Crank–Nicolson method for handling the time derivative.

8.2 Use the finite element program in Figure 8.7 to solve the one-dimensional solute transport problem for $D_L = 0$. Compare the result with the ideal step breakthrough curve.

8.3 Show by direct matrix multiplication that

$$[R][R]^{-1} = \begin{pmatrix} 1 & 0 \\ 0 & 1 \end{pmatrix}$$

8.4 Suppose that the dispersion coefficient in the flow direction is 10 cm² s⁻¹ and that, perpendicular to the flow direction, it is 1 cm² s⁻¹. If the flow direction is 30° counterclockwise to the global coordinate x axis, what is the dispersion coefficient tensor in the global coordinate system?

8.5 Write a finite difference expression for the one-dimensional solute transport equation

$$D \frac{\partial^2 C}{\partial x^2} - v \frac{\partial C}{\partial x} = \frac{\partial C}{\partial t}$$

where D is dispersion coefficient, C is concentration, and v is velocity. Have you written an explicit or an implicit approximation?

Concluding Remarks

Our intent was to bring readers to the point where it is possible to pursue two alternate (or concurrent) paths. Some readers may now want to delve into higher level texts which treat the theory of numerical methods in more depth. These readers are referred to Remson et al. (1971), Huntoon (1974), and Rushton and Redshaw (1979) for an intermediate level treatment of finite difference methods applied to subsurface flow problems, and to Pinder and Gray (1977) and Cheng (1978) for an advanced level treatment of finite element methods applied to both groundwater and surface water systems. Texts which deal with relevant numerical methods in general include those by Zienkiewicz (1977), Segerlind (1976), Huebner (1975), Cook (1974), Carnahan et al. (1969), Smith (1965), and Forsythe and Wasow (1960).

After studying our book, other readers may wish to postpone additional study of the theory of numerical methods and acquire experience with some of the general models which are readily available and well documented. Bachmat et al. (1980) present an inventory of groundwater models which have potential for use in water resource management. The two most widely used finite difference models for solving two-dimensional flow problems are those developed by Prickett and Lonnquist (1971) and Trescott et al. (1976). Both of these models use the more general form of the transient flow equation presented in Section 5.5. Thus both models allow consideration of anisotropic and heterogeneous aquifers. One of the differences between the models is the way in which T_x and T_y are defined relative to the nodal spacing. Ways of defining transmissivity in a finite difference scheme are discussed by Appel (1976).

A drawback of both the Prickett–Lonnquist model and the Trescott et al. model is that they are not always suited to modeling cross sections (profiles) where the upper boundary of the system is the water table. The difficulty centers around the fact that either the configuration of the water table is unknown, and therefore it is not possible to define the upper boundary condition, or it is a goal of the simulation to predict the change in the configuration of the water table through time. If it is known that the fluctuation of the water table will not be greater than the vertical node spacing, it is possible to specify a Neumann-type boundary condition along the upper boundary, and the model will predict the values of head at the water table. Leake (1977) modifies the Trescott et al. model to allow a moving water table in a profile model. His scheme eliminates nodes from the computation by setting nodal transmissivity values to zero as the water table drops below a given node. Leake (1977) presents an application of this modified version of the Trescott et al. model to a dewatering problem.

A moving water table in a profile model is more easily handled with the finite element method. In the finite element method, the location of each node in the array must be specified by coordinates. A change in the coordinate location of nodes positioned on the water table presents no problem, because equations are written for each node in terms of nodal coordinates. Thus coordinate locations can be updated after each iteration within a time step, or after each time step if the equations are solved directly. In this way, the finite element method allows the inclusion of deformable elements. A well-documented, two-dimensional steady-state finite element model, which allows for heterogeneous and anisotropic conditions as well as deformable elements and the presence of a seepage face, is developed by Neuman and Witherspoon (1970) and documented by Neuman (1976).

The development and testing of contaminant transport models is currently an area of active research. Several investigators have developed contaminant transport models using finite elements and, in some cases, finite difference models have also been used successfully. Konikow and Bredehoeft (1978) develop a well-documented contaminant transport model using the method of characteristics in conjunction with finite differences. The method of characteristics allows the advection-dispersion equation to be solved so that numerical errors, which in some degree plague all numerical solutions of this equation, are minimized. Konikow and Bredehoeft (1978) present the details of the method and some applications of the model. Other kinds of contaminant transport models include those in which the effects of dispersion are neglected. Anderson (1979) presents a review of the use of both types of models.

Readers may also wish to investigate other types of modeling activity which is currently receiving attention in the literature. Some recent developments are reviewed by Faust and Mercer (1980b). Appel and Bredehoeft (1976) present a review of the models being developed by the U.S. Geological Survey. More recently, models have been developed to facilitate solution of the inverse problem (Cooley, 1977 and 1979; Neuman and Yakowitz, 1979; Neuman et al., 1980; and Neuman, 1980). Other researchers are developing models in which the stochastic properties of the hydraulic conductivity distribution are taken into account (Smith and Freeze, 1979; Dagan, 1979; Bakr et al., 1978; Gutjahr et al., 1978; Gutjahr and Gelhar, 1981). Several researchers (for example, Gelhar et al., 1979; Tang and Pinder, 1979; Smith and Schwartz, 1980 and 1981) are using stochastic analysis to study contaminant transport in groundwater.

The answers generated using a mathematical model are dependent on the quality and quantity of the field data available to define the input parameters and boundary conditions. Modeling can never be a substitute for field work. Used in conjunction with good field data, a model can provide insight into the dynamics of the flow system and also serve as an invaluable predictive tool.

Anisotropy and Tensors

A.1 INTRODUCTION

When the physical properties of a medium are dependent on direction, the medium is said to be anisotropic. Darcy's law was given in Chapter 1 for the case of an isotropic medium. In this appendix, Darcy's law is generalized to a two-dimensional anisotropic medium. The mathematical description of the hydraulic conductivity as a second-rank tensor applies also to the dispersion coefficient as defined in Chapter 8.

A.2   HYDRAULIC CONDUCTIVITY TENSOR

Geologically, Figure A.1(a) represents a cross section of tilted shale beds. The coordinate system has the x axis horizontal and the y axis vertical. We call this

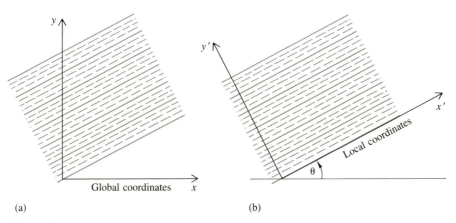

Figure A.1
Titled shale beds in (a) a global coordinate system and (b) a local coordinate system aligned with the bedding.

the global coordinate system. Suppose that groundwater can flow only parallel to the bedding, that is, the shale is impermeable perpendicular to bedding. In the global coordinate system, a head gradient in the y direction will induce some flow along the beds, and hence there will be an x component of flow. The relationship between q_x and $\partial h / \partial y$ is not covered by the form of Darcy's law introduced in Chapter 1. The generalization requires that each component of the vector $\mathbf{q}$ be linearly proportional to each component of the vector $\mathbf{grad}\ h$.

$$q_x = -K_{11} \frac{\partial h}{\partial x} - K_{12} \frac{\partial h}{\partial y} \tag{A.1a}$$

$$q_y = -K_{21} \frac{\partial h}{\partial x} - K_{22} \frac{\partial h}{\partial y} \tag{A.1b}$$

The ratio q_y / q_x is not equal to the ratio $(\partial h / \partial y)/(\partial h / \partial x)$; that is, $\mathbf{q}$ does not point in the same direction as $\mathbf{grad}\ h$.

Instead of a single proportionality constant K as in Equation 1.7, there are now four proportionality constants. The hydraulic conductivity is represented in matrix form as

$$[K] = \begin{pmatrix} K_{11} & K_{12} \\ K_{21} & K_{22} \end{pmatrix} \tag{A.2}$$

Equation A.2 is the tensor representation of hydraulic conductivity for an anisotropic medium. Hydraulic conductivity is a second-rank tensor because it relates two vectors that are first-rank tensors. A scalar quantity is a zeroth-rank tensor.

Suppose a coordinate system more natural to the geological situation were chosen. The coordinate system with the x' direction parallel to bedding and the y' direction perpendicular to bedding will be called the local coordinate system (Figure A.1b). The prime notation will be used to distinguish these coordinates from the global coordinates. In the local coordinate system, the generalized form of Darcy's law is

$$q_{x'} = -K'_{11} \frac{\partial h}{\partial x'} - K'_{12} \frac{\partial h}{\partial y'} \tag{A.3a}$$

$$q_{y'} = -K'_{21} \frac{\partial h}{\partial x'} - K'_{22} \frac{\partial h}{\partial y'} \tag{A.3b}$$

In the local coordinate system, a gradient in the y' direction will not produce flow in the x' direction, that is, $K'_{12} = 0$. Similarly, a gradient in the x' direction will not produce flow in the y' direction, and $K'_{21} = 0$. The components of the hydraulic conductivity tensor depend on the choice of coordinate system. In the coordinate system in which the off-diagonal components of the tensor are zero, the coordinate axes directions are called the principal directions.

A.3 COORDINATE SYSTEM ROTATION

Except for choice of coordinate system, the hydrogeologic problems represented in Figures A.1a and A.1b are the same. There must then be a mathematical description that transforms the problem from one set of coordinates to the other, so that the simulation of the physical situation remains unchanged. The magnitudes and directions of the two vectors, specific discharge **q** and gradient of head **grad** h, must be the same in the two coordinate systems. Only their representation in terms of components will be different in the different coordinate systems. Similarly, the components of the hydraulic conductivity tensor will be different in the two coordinate systems, but the same proportionality between **q** and **grad** h must be maintained. We use these constraints to show the mathematical transformation of the hydraulic conductivity tensor for a rotation in two dimensions.

Let θ be the counterclockwise rotation angle of the local coordinate system relative to the global coordinate system. The components of a vector **q** in the global coordinate system are related to those in the local coordinate system through the matrix equation

$$\begin{pmatrix} q_{x'} \\ q_{y'} \end{pmatrix} = [R] \begin{pmatrix} q_x \\ q_y \end{pmatrix} \tag{A.4}$$

where the rotation matrix $[R]$ is defined by

$$[R] = \begin{pmatrix} \cos\theta & \sin\theta \\ -\sin\theta & \cos\theta \end{pmatrix} \tag{A.5}$$

The rotation matrix is described in most texts on vector algebra. Just as in Equation A.4, the components of the gradient vector in the global coordinate system are related to those in the local coordinate system through the matrix equation

$$\begin{pmatrix} \dfrac{\partial h}{\partial x'} \\ \dfrac{\partial h}{\partial y'} \end{pmatrix} = [R] \begin{pmatrix} \dfrac{\partial h}{\partial x} \\ \dfrac{\partial h}{\partial y} \end{pmatrix} \tag{A.6}$$

Equation A.3, Darcy's law, can be written in matrix form also.

$$\begin{pmatrix} q_{x'} \\ q_{y'} \end{pmatrix} = -[K'] \begin{pmatrix} \dfrac{\partial h}{\partial x'} \\ \dfrac{\partial h}{\partial y'} \end{pmatrix} \tag{A.7}$$

Substituting Equations A.4 and A.6 into Equation A.7 and multiplying across by the inverse of $[R]$ gives

$$\begin{pmatrix} q_x \\ q_y \end{pmatrix} = -[R]^{-1}[K'][R] \begin{pmatrix} \dfrac{\partial h}{\partial x} \\ \dfrac{\partial h}{\partial y} \end{pmatrix} \tag{A.8}$$

The inverse rotation matrix $[R]^{-1}$ is obtained by substituting $-\theta$ for θ. It rotates a vector from the local coordinate system back into the global system. The $[R]^{-1}$ matrix is

$$[R]^{-1} = \begin{pmatrix} \cos\theta & -\sin\theta \\ \sin\theta & \cos\theta \end{pmatrix} \tag{A.9}$$

Equation A.8 is the matrix form of Darcy's law in global coordinates when we make the identification

$$[K] = [R]^{-1}[K'][R] \tag{A.10}$$

Equation A.10 is the basic result of this appendix. It describes how the hydraulic conductivity tensor components transform with a coordinate rotation.

In most flow problems, it is possible to define the global coordinate system to coincide with the principal directions of the hydraulic conductivity tensor. In this case, one need only define K_{11} and K_{22}, because K_{12} and K_{21} will be zero. If it is not possible to define such a global coordinate system, it will be necessary to work with both local and global coordinate systems. Equation A.10 is used to find K_{11}, K_{22}, K_{12}, and K_{21} for each node or element, given that in the local coordinate system $K'_{12} = K'_{21} = 0$, and K'_{11} and K'_{22} are supplied from field data.

In contaminant transport problems where the flow field is not uniform, the global and local coordinate systems do not coincide in general. In this case, the dispersion coefficient tensor must be rotated between the coordinate systems as necessary throughout the problem domain according to Equation 8.16.

Variational Method

B.1 INTRODUCTION

The variational method is an alternate formulation for problems that can be posed in terms of an extremum principle. In this appendix, we show that the variational approach for steady-state groundwater flow leads to the same algebraic equations as the Galerkin method.

B.2 MINIMUM DISSIPATION PRINCIPLE

As groundwater flows, it gives up its potential energy to friction. The dissipation of the groundwater system is defined to be one-half the rate of energy loss. The integral expression for the dissipation can be derived from the physics of the flow. The extremum principle that governs the problem is that the dissipation, subject to boundary conditions, be minimum.

The potential energy per unit mass of water is

$$\phi(x, y) = gh(x, y) \tag{B.1}$$

The potential ϕ is a force potential

$$F_x = -\frac{\partial \phi}{\partial x} \quad \text{and} \quad F_y = -\frac{\partial \phi}{\partial y} \tag{B.2}$$

where F_x and F_y are the components of the force per unit mass of water. The force components per unit volume of aquifer are $n\rho_w F_x$ and $n\rho_w F_y$, where n is the porosity and ρ_w is the density of water.

The average linear velocity is $\mathbf{q}/n$, where $\mathbf{q}$ is the Darcy velocity. At each point in the groundwater flow system, the rate of energy loss per unit of aquifer volume is the scalar product of the force per unit of aquifer volume times the average linear velocity, that is, $\rho_w(F_x q_x + F_y q_y)$. This expression must be integrated over the aquifer volume. If Equations B.1, B.2, and Darcy's law are substituted, then the dissipation J that is defined to be one-half the total rate of energy loss is

$$J = \frac{\rho_w g b}{2} \iint_D \left\{ K \left(\frac{\partial h}{\partial x} \right)^2 + K \left(\frac{\partial h}{\partial y} \right)^2 \right\} dx\, dy \tag{B.3}$$

where b is the aquifer thickness.

The expression for the dissipation in Equation B.3 was derived in accordance with the basic laws of physics and has units of energy per unit time. If we eliminate the multiplicative constants, the expression in Equation B.3 becomes

$$J = \frac{1}{2} \iint_D \left\{ \left(\frac{\partial h}{\partial x} \right)^2 + \left(\frac{\partial h}{\partial y} \right)^2 \right\} dx\, dy \tag{B.4}$$

Minimization of the expression in Equation B.4 is equivalent to the minimization of the dissipation in Equation B.3 because the two expressions are simple multiples of each other.

If the boundary heads are specified or if no-flow conditions are specified, then Equation B.4 is the total variational integral which must be minimized. If some nonzero boundary flows are specified, an additional boundary integral term must be included in Equation B.4. In the next section, we assume that Equation B.4 applies.

B.3 FINITE ELEMENTS

The subdivision of the problem domain into finite elements and the definition of element nodal basis functions are considered to be done just as in Chapter 6. The nodal basis functions that multiply the nodal heads define a trial solution $\hat{h}(x, y)$ which is piecewise continuous throughout the problem domain.

The nodal heads h_L which minimize the integral J constitute the numerical solution. By the finite element approximation, the integral J can be considered to be a function of the many variables h_L, $L = 1, 2, \ldots, NNODE$, where $NNODE$ is the number of nodes. An extremum value of J is found from the requirement that its partial derivative with respect to each h_L is equal to zero; that is, $\partial J/\partial h_L = 0$ for each L. This condition produces one algebraic equation per node just as the Galerkin method does.

We consider the integral J to be evaluated element by element. The contribution to the total value of J from element e is

$$J^e = \frac{1}{2} \iint_e \left\{ \left(\frac{\partial \hat{h}^e}{\partial x} \right)^2 + \left(\frac{\partial \hat{h}^e}{\partial y} \right)^2 \right\} dx\, dy \tag{B.5}$$

If Equations 6.14a and 6.14b are substituted for $\partial \hat{h}^e/\partial x$ and $\partial \hat{h}^e/\partial y$, then

$$J^e = \frac{1}{2} \iint_e \left\{ \left(\frac{\partial N_i^e}{\partial x} h_i + \frac{\partial N_j^e}{\partial x} h_j + \frac{\partial N_m^e}{\partial x} h_m \right)^2 \right.$$
$$\left. + \left(\frac{\partial N_i^e}{\partial y} h_i + \frac{\partial N_j^e}{\partial y} h_j + \frac{\partial N_m^e}{\partial y} h_m \right)^2 \right\} dx\, dy \tag{B.6}$$

The contribution to $\partial J/\partial h_L$ from the element e is by the chain rule of differentiation

$$\frac{\partial J^e}{\partial h_L} = \iint_e \left\{ \left(\frac{\partial N_i^e}{\partial x} h_i + \frac{\partial N_j^e}{\partial x} h_j + \frac{\partial N_m^e}{\partial x} h_m \right) \frac{\partial N_L^e}{\partial x} \right.$$
$$\left. + \left(\frac{\partial N_i^e}{\partial y} h_i + \frac{\partial N_j^e}{\partial y} h_j + \frac{\partial N_m^e}{\partial y} h_m \right) \frac{\partial N_L^e}{\partial y} \right\} dx\, dy \tag{B.7}$$

where $L = i, j,$ or m. Equation B.7 is the element contribution to the left-hand side of the Lth equation. Equation B.7 in identical to the right side of Equation (6.18) for $L = i, j,$ or m. Hence we have now arrived at the same point by the variational and Galerkin methods.

Isoparametric Quadrilateral Elements

C.1   INTRODUCTION

Quadrilateral elements can be applied to an irregular problem geometry in much the same way as triangular elements. The sides of a quadrilateral can linearly approximate a curved boundary. Because the sides of a quadrilateral are inclined to the coordinate axes, basis functions with an xy term will not be linear along element boundaries, and they will not be continuous over element boundaries.

   The solution to the problem of defining basis functions is to use a one-to-one mapping of a square onto the quadrilateral through a coordinate transforma-

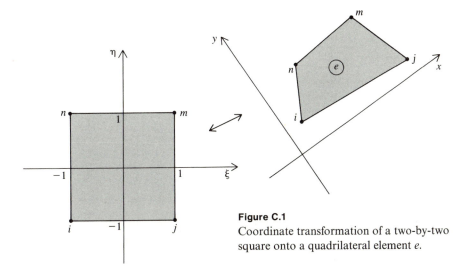

Figure C.1
Coordinate transformation of a two-by-two
square onto a quadrilateral element e.

tion (Figure C.1). The basis functions developed in Chapter 7 for a rectangular element are carried over to the quadrilateral element by the mapping. The quadrilateral in the (x, y) coordinates of the problem domain is defined in terms of the (ξ, η) coordinates of the square such that corner nodes of the quadrilateral correspond to corner nodes of the square.

The coordinates $x = x(\xi, \eta)$ and $y = y(\xi, \eta)$ are functions of the auxiliary variables ξ and η. If the trial solution $\hat{h}^e(\xi, \eta)$ is defined over the (ξ, η) square in terms of the nodal values at the corners, then $\hat{h}^e(x, y)$ is defined in the problem domain coordinates by combining the function $\hat{h}^e(\xi, \eta)$ with the coordinate transformation equations. This composite definition of $\hat{h}^e(x, y)$ is called a parametric representation of the function. The quadrilateral element is called isoparametric because the basis functions that define the trial solution over the square also define the coordinate transformation.

C.2 COORDINATE TRANSFORMATION

The purpose of the coordinate transformation is to distort the archetypal square, $-1 \le \xi \le 1, -1 \le \eta \le 1$, in (ξ, η) coordinates, into the corresponding quadrilateral in (x, y) coordinates. The element basis functions $\tilde{N}^e_L(\xi, \eta)$ defined in Chapter 7, Equation 7.16, can be used to interpolate the corner values, including the coordinates themselves, over the square.

$$x = x(\xi, \eta) = \tilde{N}_i^e(\xi, \eta)x_i + \tilde{N}_j^e(\xi, \eta)x_j + \tilde{N}_m^e(\xi, \eta)x_m + \tilde{N}_n^e(\xi, \eta)x_n \quad \text{(C.1a)}$$

$$y = y(\xi, \eta) = \tilde{N}_i^e(\xi, \eta)x_i + \tilde{N}_j^e(\xi, \eta)x_j + \tilde{N}_m^e(\xi, \eta)x_m + \tilde{N}_n^e(\xi, \eta)x_n \quad \text{(C.1b)}$$

where

$$\tilde{N}_i^e(\xi, \eta) = \tfrac{1}{4}(1 - \xi)(1 - \eta) \qquad \text{(C.2a)}$$

$$\tilde{N}_j^e(\xi, \eta) = \tfrac{1}{4}(1 + \xi)(1 - \eta) \qquad \text{(C.2b)}$$

$$\tilde{N}_m^e(\xi, \eta) = \tfrac{1}{4}(1 + \xi)(1 + \eta) \qquad \text{(C.2c)}$$

$$\tilde{N}_n^e(\xi, \eta) = \tfrac{1}{4}(1 - \xi)(1 + \eta) \qquad \text{(C.2d)}$$

Observe, as a check, that these equations provide the result that the corners of the square $(-1, -1)$, $(1, -1)$, $(1, 1)$, and $(-1, 1)$ correspond, respectively, to the corners of the quadrilateral (x_i, y_i), (x_j, y_j), (x_m, y_m), and (x_n, y_n).

Equation C.1 enables us, in theory, to go back and forth from (x, y) to (ξ, η) coordinates. The transformation of coordinates is analogous to the transformation between Cartesian (x, y) and polar (r, ϕ) coordinates.

Element Trial Solution

The trial solution $\hat{h}^e(x, y)$ over the quadrilateral in (x, y) space is implicitly defined by $\hat{h}^e(\xi, \eta)$ over the square in (ξ, η) space

$$\hat{h}^e(\xi, \eta) = \tilde{N}_i^e(\xi, \eta)h_i + \tilde{N}_j^e(\xi, \eta)h_j + \tilde{N}_m^e(\xi, \eta)h_m + \tilde{N}_n^e(\xi, \eta)h_n \quad \text{(C.3)}$$

where the nodal values h_i, h_j, h_m, and h_n are the nodal values at the corners of the square, and hence the quadrilateral. The basis functions $\tilde{N}_i^e(\xi, \eta)$, $\tilde{N}_j^e(\xi, \eta)$, $\tilde{N}_m^e(\xi, \eta)$, and $\tilde{N}_n^e(\xi, \eta)$ serve both as mapping functions for the coordinate transformation and as interpolation functions for the nodal values of head. The basis functions are called shape functions for their coordinate mapping role, and they are called interpolation functions for their interpolating role. The trial solution $\hat{h}^e(x, y)$ is now defined parametrically over the archetypal quadrilateral.

Transformation of Integrals

The composite definition of the quadrilateral element basis functions means that the (x, y) integrations in element matrices are most conveniently done as

(ξ, η) integrations over the square parent element. The conductance matrix term $G_{L,i}^e$ is used as an example to show the transformation procedure.

$$G_{L,i}^e = \iint_{\text{Quad}} \left(\frac{\partial N_i^e}{\partial x} \frac{\partial N_L^e}{\partial x} + \frac{\partial N_i^e}{\partial y} \frac{\partial N_L^e}{\partial y} \right) dx\,dy \tag{C.4}$$

where the integration is over the quadrilateral element and the basis functions N_i^e and N_L^e are composite functions of x and y. This integral needs to be transformed to one over the (ξ, η) square. The transformation of a multiple integral is described in most books on advanced calculus. Two aspects are involved in the transformation. The derivatives in the integrand must be evaluated by the chain rule of differentiation, and the area element $dx\,dy$ must be expressed as a scalar multiple of $d\xi\,d\eta$.

The Jacobian matrix of the transformation from (ξ, η) to (x, y) coordinates is defined to be

$$[J] = \begin{pmatrix} \dfrac{\partial x}{\partial \xi} & \dfrac{\partial y}{\partial \xi} \\[2ex] \dfrac{\partial x}{\partial \eta} & \dfrac{\partial y}{\partial \eta} \end{pmatrix} \tag{C.5}$$

The Jacobian matrix represents the chain rule of differentiation for composite functions.

$$\begin{pmatrix} \dfrac{\partial \tilde{N}_L^e}{\partial \xi} \\[2ex] \dfrac{\partial \tilde{N}_L^e}{\partial \eta} \end{pmatrix} = [J] \begin{pmatrix} \dfrac{\partial N_L^e}{\partial x} \\[2ex] \dfrac{\partial N_L^e}{\partial y} \end{pmatrix} \tag{C.6}$$

The four derivatives in $[J]$ can be written explicitly from the definition of the coordinate transformation, Equation C.1.

The inverse of $[J]$ is the Jacobian matrix of the inverse transformation from (x, y) to (ξ, η) coordinates.

$$[J]^{-1} = \begin{pmatrix} \dfrac{\partial \xi}{\partial x} & \dfrac{\partial \eta}{\partial x} \\[2ex] \dfrac{\partial \xi}{\partial y} & \dfrac{\partial \eta}{\partial y} \end{pmatrix} \tag{C.7}$$

The entries in the inverse of $[J]$ can be written explicitly in terms of the entries of $[J]$ itself.

$$\frac{\partial \xi}{\partial x} = \frac{1}{|J|} \frac{\partial y}{\partial \eta} \tag{C.8a}$$

$$\frac{\partial \xi}{\partial y} = -\frac{1}{|J|} \frac{\partial y}{\partial \xi} \tag{C.8b}$$

$$\frac{\partial \eta}{\partial x} = -\frac{1}{|J|} \frac{\partial x}{\partial \eta} \tag{C.8c}$$

$$\frac{\partial \eta}{\partial y} = \frac{1}{|J|} \frac{\partial x}{\partial \xi} \tag{C.8d}$$

where

$$|J| = \frac{\partial x}{\partial \xi} \frac{\partial y}{\partial \eta} - \frac{\partial y}{\partial \xi} \frac{\partial x}{\partial \eta} \tag{C.9}$$

The inverse Jacobian matrix represents the chain rule

$$\begin{pmatrix} \dfrac{\partial N_L^e}{\partial x} \\ \dfrac{\partial N_L^e}{\partial y} \end{pmatrix} = [J]^{-1} \begin{pmatrix} \dfrac{\partial \tilde{N}_L^e}{\partial \xi} \\ \dfrac{\partial \tilde{N}_L^e}{\partial \eta} \end{pmatrix} \tag{C.10}$$

that is,

$$\frac{\partial N_L^e}{\partial x} = \frac{\partial \tilde{N}_L^e}{\partial \xi} \frac{\partial \xi}{\partial x} + \frac{\partial \tilde{N}_L^e}{\partial \eta} \frac{\partial \eta}{\partial x} \tag{C.11a}$$

$$\frac{\partial N_L^e}{\partial y} = \frac{\partial \tilde{N}_L^e}{\partial \xi} \frac{\partial \xi}{\partial y} + \frac{\partial \tilde{N}_L^e}{\partial \eta} \frac{\partial \eta}{\partial y} \tag{C.11b}$$

The derivatives that appear in Equation C.4 can now be expressed in (ξ, η) coordinates through Equations C.8, C.9, and C.11.

The other requirement for the transformation of the integral is that a differential area $dx\,dy$ in the quadrilateral corresponds to $|J|\,d\xi\,d\eta$ in the square. The scale factor $|J|$ is a function of position within the (ξ, η) square.

In summary, the transformed integral is

$$
G^e_{L,i} = \int_{-1}^{-1} \int_{-1}^{-1} \left\{ \left(\frac{\partial \tilde{N}^e_i}{\partial \xi} \frac{\partial \xi}{\partial x} + \frac{\partial \tilde{N}^e_i}{\partial \eta} \frac{\partial \eta}{\partial x} \right) \left(\frac{\partial \tilde{N}^e_L}{\partial \xi} \frac{\partial \xi}{\partial x} + \frac{\partial \tilde{N}^e_L}{\partial \eta} \frac{\partial \eta}{\partial x} \right) \right.
$$
$$
\left. + \left(\frac{\partial \tilde{N}^e_i}{\partial \xi} \frac{\partial \xi}{\partial y} + \frac{\partial \tilde{N}^e_i}{\partial \eta} \frac{\partial \eta}{\partial y} \right) \left(\frac{\partial \tilde{N}^e_L}{\partial \xi} \frac{\partial \xi}{\partial y} + \frac{\partial \tilde{N}^e_L}{\partial \eta} \frac{\partial \eta}{\partial y} \right) \right\} |J|\,d\xi\,d\eta \quad \text{(C.12)}
$$

C.3 COMPUTER PROGRAM MODIFICATION

The coordinate transformation can be incorporated in a straightforward manner in a computer code in which the ξ and η integrations, such as Equation C.12, are done by evaluating the integrands at the four Gauss points within the square. The modification of the loop *DO 30 IQ*$=1,4$ of the program in Figure 7.3 is shown in Figure C.2 for the conductance matrix. First, the ξ and η derivatives of the basis functions $\tilde{N}^e_L(\xi, \eta)$ are evaluated. Second, the Jacobian and its inverse are computed. Third, the chain rule expressions for $\partial N_L/\partial x$ and $\partial N_L/\partial y$ are computed. Finally, the contributions to the conductance matrix for that Gauss point are added.

Figure C.2

Element conductance matrix terms computed by Gaussian quadrature for isoparametric quadrilateral elements.

```
C    GAUSSIAN QUADRATURE
     DO 30 IQ=1,4
     NXSI(1)=-.25*(1-ETA(IQ))
     NXSI(2)=.25*(1-ETA(IQ))
     NXSI(3)=.25*(1+ETA(IQ))
     NXSI(4)=-.25*(1+ETA(IQ))
     NETA(1)=-.25*(1-XSI(IQ))
     NETA(2)=-.25*(1+XSI(IQ))
     NETA(3)=.25*(1+XSI(IQ))
     NETA(4)=.25*(1-XSI(IQ))
C    DERIVATIVES OF TRANSFORMATION (X,Y) TO (XSI,ETA)
     XXSI=NXSI(1)*X(I)+NXSI(2)*X(J)+NXSI(3)*X(M)+NXSI(4)*X(N)
     XETA=NETA(1)*X(I)+NETA(2)*X(J)+NETA(3)*X(M)+NETA(4)*X(N)
     YXSI=NXSI(1)*Y(I)+NXSI(2)*Y(J)+NXSI(3)*Y(M)+NXSI(4)*Y(N)
     YETA=NETA(1)*Y(I)+NETA(2)*Y(J)+NETA(3)*Y(M)+NETA(4)*Y(N)
     DET=XXSI*YETA-YXSI*XETA
C    DERIVATIVES OF INVERSE TRANSFORMATION (XSI,ETA) TO (X,Y)
     XSIX=YETA/DET
     XSIY=-YXSI/DET
     ETAX=-XETA/DET
     ETAY=XXSI/DET
     DO 25 KKK=1,4
     NX(KKK)=NXSI(KKK)*XSIX+NETA(KKK)*ETAX
     NY(KKK)=NXSI(KKK)*XSIY+NETA(KKK)*ETAY
25   CONTINUE
     G(L,I)=G(L,I)+DET*(NX(1)*NX(KK)+NY(1)*NY(KK))
     G(L,J)=G(L,J)+DET*(NX(2)*NX(KK)+NY(2)*NY(KK))
     G(L,M)=G(L,M)+DET*(NX(3)*NX(KK)+NY(3)*NY(KK))
     G(L,N)=G(L,N)+DET*(NX(4)*NX(KK)+NY(4)*NY(KK))
30   CONTINUE
```

Analogies

D.1 INTRODUCTION

Physical analogies have played an important role in the development of mathematical modeling of groundwater flow. If the mathematical equations that represent two physical situations are equivalent, then the solution techniques and solutions themselves are applicable to both problems. In this appendix, three analogies to groundwater flow are described. The groundwater variables and their electrical and heat flow counterparts are summarized in Table D.1. Table D.2 is a similar summary of the structural mechanics analogy.

D.2 ELECTRICAL ANALOGY

M. King Hubbert's interest in the physics of groundwater flow was aroused while he was conducting electrical earth-resistivity surveys during the period from 1931 to 1936. In a resistivity survey, four electrodes are driven into the ground. A battery produces a current I through the two outer electrodes. The voltage difference V is measured across the two inner electrodes. If the sub-

Table D.1

Analogies to groundwater flow (representative units are shown in brackets)

VARIABLE	GROUNDWATER	ELECTRICITY	HEAT
Potential	Head, h [cm]	Voltage, V [Volts]	Temperature, T [°C]
Quantity transported	Volume discharge rate [$cm^3 s^{-1}$]	Electrical charge [Coulomb]	Heat [calorie]
Physical property of medium	Hydraulic conductivity, K [$cm\ s^{-1}$]	Electrical conductivity, σ [mhos m^{-1}]	Thermal conductivity, K [cal cm^{-1} s^{-1} °C^{-1}]
Relation between potential and flow field	Darcy's law $\mathbf{q} = -K\ \mathbf{grad}\ h$ where $\mathbf{q}$ is specific discharge [$cm\ s^{-1}$]	Ohm's law $\mathbf{i} = -\sigma\ \mathbf{grad}\ V$ where $\mathbf{i}$ is electrical current [Amperes]	Fourier's law $\mathbf{q} = -K\ \mathbf{grad}\ T$ where $\mathbf{q}$ is heat flow [cal cm^{-2} s^{-1}]
Storage quantity	Specific storage, S_s [cm^{-1}]	Capacitance, C [microfarad]	Heat capacity, C_v [cal cm^{-3} °C^{-1}]

surface of the earth is electrically uniform, then the theoretical lines of electrical current flow are shown by the dashed lines of Figure D.1. Perpendicular to these flow lines are the solid lines of constant voltage, that is, equipotential lines.

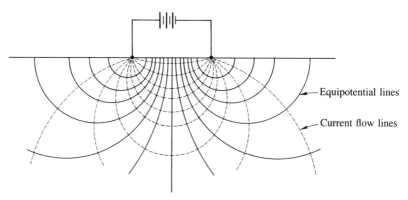

Figure D.1

Experimental set-up for conducting an earth-resistivity survey. (After Dobrin, 1976.)

Table D.2
Structural mechanics analogy to groundwater flow

VARIABLE	GROUNDWATER	MECHANICS
Unknown variable	Head, h	Displacement, $\delta = \begin{pmatrix} u \\ v \end{pmatrix}$
First derivative quantity	$\mathbf{grad}\ h = \begin{pmatrix} \dfrac{\partial h}{\partial x} \\ \dfrac{\partial h}{\partial y} \end{pmatrix}$	Strain, $\varepsilon = \begin{pmatrix} \dfrac{\partial u}{\partial x} \\ \dfrac{\partial v}{\partial y} \\ \dfrac{\partial u}{\partial y} + \dfrac{\partial v}{\partial x} \end{pmatrix}$
Conjugate variable	$\mathbf{q} = \begin{pmatrix} q_x \\ q_y \end{pmatrix}$	Stress, $\sigma = \begin{pmatrix} \sigma_{xx} \\ \sigma_{yy} \\ \tau_{xy} \end{pmatrix}$
Physical property of medium	Hydraulic conductivity, $$[K] = \begin{pmatrix} K & 0 \\ 0 & K \end{pmatrix}$$	Elastic moduli, $$[C] = \begin{pmatrix} \dfrac{E}{1-v^2} & \dfrac{v}{1-v^2} & 0 \\ \dfrac{v}{1-v^2} & \dfrac{E}{1-v^2} & 0 \\ 0 & 0 & G \end{pmatrix}$$ where E is Young's modulus, v is Poisson's ratio, and G is shear modulus, $G = \dfrac{E}{2(1+v)}$
Assembled finite element matrix equation	$[G]\{h\} = \{B\}$ where $[G]$ is conductance matrix and $\{B\}$ is recharge matrix	$[S]\{\delta\} = \{F\}$ where $[S]$ is stiffness matrix and $\{F\}$ is load matrix

Hubbert (1969, p. 11) asks the question: "Suppose that the ground consisted of a uniform sand covered with an impermeable layer and filled with water. Suppose further that the electrodes were to be replaced by wells terminated at the top of the sand by hemispherical screens, and the electrical battery replaced by a pump connecting the two wells. Then, with the pump operating at a uniform rate, a water flow field would be created through the sand between the two wells. What would be the nature of the flow field, and what would be the equations describing the flow?"

Hubbert intuitively felt that the water flow lines would be the same as the electrical current flow lines. The analogy is exact in the sense that mathematically identical equations describe both cases. For a particular material, the current I is directly proportional to voltage drop $V_2 - V_1$ and cross-sectional area A, but inversely proportional to length $\ell_2 - \ell_1$. Calling the proportionality constant the electrical conductivity σ gives the relationship

$$I = -\sigma A \frac{V_2 - V_1}{\ell_2 - \ell_1} \tag{D.1}$$

Equation D.1 is Ohm's law. A more common form is $I = (V_2 - V_1)/R$, where the resistance $R = (\ell_2 - \ell_1)/\sigma A$.

Ohm's law and Darcy's law are exactly analogous mathematically. They are experimental laws, they relate a flow to a potential, and they contain a constant of proportionality that is a material property.

The use of capacitors in an electrical analog model permits simulation of time dependent problems. A capacitor stores or releases charge when the voltage across its plates changes. Capacitance is defined as the amount of charge required to raise the voltage across the plates one unit. It is equivalent to the specific storage of an aquifer. Electrical analog models and their relationship to finite difference models are described in detail by Rushton and Redshaw (1979) and by Bennett (1976).

D.3 HEAT FLOW ANALOGY

C. V. Theis (1935) developed an equation for the time-dependent radial flow to a well by translating the heat flow analogy into groundwater terms. Heat flows by conduction in solids from regions of higher temperature to regions of lower temperature. The heat flow is proportional to the temperature gradient

$$q = -K \frac{dT}{d\ell} \tag{D.2}$$

where q is the quantity of heat crossing a unit area in unit time, T is temperature, ℓ is distance, and K is the thermal conductivity. Equation D.2 is Fourier's law. Heat flow is analogous to specific discharge, temperature is analogous to head,

and thermal conductivity is analogous to hydraulic conductivity. Both conductivities are properties of the medium.

The island recharge problem of Section 3.4 can be translated into the following
heat flow problem. Consider a glass warming plate whose length is twice its
width and whose thickness is b. If the boundaries are kept at 0°C, what heating
rate (calories per square centimeter per second) is required to keep the center
of the tray at 20°C? Now, suppose the heating element is turned off. The tray
continues to stay warm, cooling down only gradually, for the same reason that
well levels gradually drop during a drought. The glass stores heat as the temperature rises, and it releases heat as the temperature falls. The measure of this
property is the heat capacity, which is defined as the heat required to raise a
unit volume a unit degree of temperature.

Carslaw and Jaeger (1959) is the standard treatise on problems of heat conduction. The analytical solutions to the island recharge problem and the reservoir lowering problem in Chapter 4 are translated from Carslaw and Jaeger. An
older but very readable text on problems of heat conduction is Ingersoll, Zobel,
and Ingersoll (1954).

D.4   STRUCTURAL MECHANICS ANALOGY

The finite element method has its origin in problems of structural mechanics.
The groundwater hydrologist can borrow from the extensive literature, including textbooks and computer programs, on establishing the mathematical equivalence between stress-strain analysis and groundwater flow. The computer
programs are generally applicable to solving groundwater problems involving
Laplace's equation or Poisson's equation. The position of a bending beam is
analogous to the position of a water table. The reader is referred to Zienkiewicz
(1977) or Cook (1974) as examples of finite element texts based on stress-strain
problems.

The correspondence between groundwater variables and their solid mechanics counterparts is more complicated and artificial than in the electrical
and heat conduction analogies. The basic quantities dealt with in the mechanics
of solids are stress σ and strain ε. Both stress and strain are second-rank tensors
that can be expressed as column matrices. They are related by Hooke's law

$$\{\sigma\} = [C]\{\varepsilon\} \tag{D.3}$$

where $[C]$ is a fourth-rank tensor of elastic moduli.

Hooke's law plays the role of Darcy's law in the analogy if the elastic modulus tensor is interpreted properly. Some care needs to be exercised in setting up the analogy between hydraulic conductivity and the elastic moduli. Only two elastic moduli are independent for an isotropic medium. Therefore, if Poisson's ratio v is set to zero and if Young's modulus E is set to hydraulic conductivity K, then it may be necessary to force shear modulus G to equal E in the stress-strain program. Also, the $[C]$ matrix in Table D.2 is for the plane stress case. It has a different form for the plane strain case.

In the mechanics problem, the unknown variables at each point in the problem domain are the horizontal component $u(x, y)$ and the vertical component $v(x, y)$ of the displacement. By fixing v to be zero at all nodal points, the horizontal displacement u is equivalent to head h. In two dimensions, there are three components of strain and stress (Table D.2). The components $\varepsilon_{xx} = \partial u/\partial x$ and $\varepsilon_{yy} = \partial v/\partial y$ are longitudinal strains, and $\varepsilon_{xy} = \partial u/\partial y + \partial v/\partial x$ is the shear strain. Similarly, σ_{xx} and σ_{yy} are normal stresses and τ_{xy} is the shear stress. The strain and stress are equivalent to **grad** h and **q**, respectively, if the second component of the strain and stress are ignored.

In the matrix equation that represents the finite element solution of the stress-strain problem, the assembled global stiffness matrix $[S]$ is analogous to the conductance matrix $[G]$, and the load matrix $\{F\}$ is analogous to the recharge matrix $\{B\}$. The load matrix can handle either point loads (that is, wells) or distributed loads (that is, recharge).

Another analogy is expressed in Equation D.3. The elastic modulus tensor is exactly analogous in its properties to the dispersivity tensor (Bear, 1961). The dispersivity tensor multiplied by the second-rank velocity tensor yields the second-rank dispersion coefficient tensor.

Glossary of Symbols

Dimensions (M, mass; L, distance; T, time) are given in brackets.

a_L longitudinal dispersivity $[L]$

a_T transverse dispersivity $[L]$

b aquifer thickness $[L]$

C solute concentration $[ML^{-3}]$

D_L longitudinal dispersion coefficient $[L^2 T^{-1}]$

D_T transverse dispersion coefficient $[L^2 T^{-1}]$

g acceleration of gravity $[LT^{-2}]$

h head $[L]$

$\hat{h}$ trial solution for head $[L]$

(i, j) (column, row) index in finite difference grid

K hydraulic conductivity $[LT^{-1}]$

m iteration index

n time step index (as a superscript)

n porosity $[L^3 L^{-3}]$

$\hat{\mathbf{n}}$ unit vector normal to curve

N_L Lth nodal basis function

P pressure $[MLT^{-2}L^{-2}]$

q	specific discharge; Darcy velocity $[LT^{-1}]$
Q	volumetric discharge rate $[L^3T^{-1}]$
r	radius $[L]$
R	recharge rate $[LT^{-1}]$
$[R]$	rotation matrix
S	storage coefficient $[L^3L^{-3}]$
t	time $[T]$
T	transmissivity $[L^2T^{-1}]$
v	pore velocity; average linear velocity $[LT^{-1}]$
$W(u)$	well function
(x, y, z)	Cartesian coordinates $[L]$
α	time step weighting parameter
ρ_w	density of water $[ML^{-3}]$
ϕ	groundwater potential (Hubbert's force potential) $[L^2T^{-2}]$
ω	relaxation parameter in successive over relaxation

References

Anderson, M. P. 1979. "Using Models to Simulate the Movement of Contaminants Through Groundwater Flow Systems." *Critical Reviews in Environmental Control* 9(2):97–156.

Andrews, C. B. 1978. "The Impact of the Use of Heat Pumps on Ground-Water Temperatures." *Ground Water* 16(6):437–443.

Andrews, C. B., and M. P. Anderson. 1979. "Thermal Alteration of Groundwater Caused by Seepage from a Cooling Lake." *Water Resources Research* 15(3):595–602.

Appel, C. A. 1976. "A Note on Computing Finite Difference Interblock Transmissivities." *Water Resources Research* 12(3):561–563.

Appel, C. A., and J. D. Bredehoeft. 1976. *Status of Ground-Water Modeling in the U.S. Geological Survey*. (U.S. Geol. Survey, Circular 737.) 9 pp.

Bachmat, Y., et al. (eds.). 1980. *Groundwater Management: The Use of Numerical Models*. (Amer. Geophysical Union, Water Resources Monograph Series-5.) 127 pp.

Back, W., and J. A. Cherry. 1976. "Chemical Aspects of Present and Future Hydrogeologic Problems." In *Advances in Groundwater Hydrology*, Z. A. Saleem, ed. Minneapolis: American Water Resources Association, 153–172.

Bakr, A. A., et al. 1978. "Stochastic Analysis of Spatial Variability in Subsurface Flows, 1: Comparison of One- and Three-Dimensional Flows." *Water Resources Research* 14(2):263–271.

Bear, J. 1961. "On the Tensor Form of Dispersion in Porous Media." *Journal of Geophysical Research* 66(4):1185–1197.

Bear, J. 1972. *Dynamics of Fluids in Porous Media*. New York: American Elsevier, 764 pp.

Bennett, G. D. 1976. *Introduction to Ground-Water Hydraulics: A Programed Text for Self-Instruction*. (U.S. Geol. Survey, Techniques of Water-Resources Investigations, Book 3, Chapter B2.) 172 pp.

Brebbia, C. A., and J. J. Connor. 1977. *Finite Element Techniques for Fluid Flow*. London, Boston: Newnes-Butterworths, 310 pp.

Bredehoeft, J. D., and G. F. Pinder. 1973. "Mass Transport in Flowing Groundwater." *Water Resources Research* 9(1):194–210.

Carnahan, B., H. A. Luther, and J. O. Wilkes. 1969. *Applied Numerical Methods*. New York: Wiley, 604 pp.

Carslaw, H. S., and J. C. Jaeger. 1959. *Conduction of Heat in Solids* (2nd ed.). London: Oxford University Press, 510 pp.

Cheng, R. T. 1978. "Modeling of Hydraulic Systems by Finite-Element Methods." In *Advances in Hydroscience*, Vol. 11. New York: Academic Press, 207–284.

Cherry, J. A., R. W. Gillham, and J. F. Pickens. 1975. "Contaminant Hydrogeology, Part 1: Physical Processes." *Geoscience Canada* 2(1):76–84.

Childs, E. C., and E. G. Youngs. 1968. "Note on the Paper 'Explanation of Paradoxes in Dupuit-Forchheimer Seepage Theory,' by Don Kirkham." *Water Resources Research* 4(1):219–220, with reply by Don Kirkham, 221–222.

Collins, M. A., L. W. Gelhar, and J. L. Wilson. 1972. "Hele–Shaw Model of Long Island Aquifer System." *Journal of the Hydraulics Division* (American Society of Civil Engineers) 98(HY9):1701–1714.

Cook, R. D. 1974. *Concepts and Applications of Finite Element Analysis: A Treatment of the Finite Element Method as Used for the Analysis of Displacement, Strain, and Stress*. New York: Wiley, 402 pp.

Cooley, R. L. 1977. "A Method of Estimating Parameters and Assessing Reliability for Models of Steady State Groundwater Flow, 1: Theory and Numerical Properties." *Water Resources Research* 13(2):318–324.

Cooley, R. L. 1979. "A Method of Estimating Parameters and Assessing Reliability for Models of Steady State Groundwater Flow, 2: Application of Statistical Analysis." *Water Resources Research* 15(3):603–617.

Dagan, G. 1979. "Models of Groundwater Flow in Statistically Homogeneous Porous Formations." *Water Resources Research* 15(1):47–63.

De Marsily, G., et al. 1977. "Nuclear Waste Disposal: Can the Geologist Guarantee Isolation?" *Science* 197(4303):519–527.

Desai, C. S. 1979. *Elementary Finite Element Method*. Englewood Cliffs, N.J.: Prentice-Hall, 434 pp.

Dobrin, M. 1976. *Introduction to Geophysical Prospecting* (3rd ed.). Hightstown, N.J.: McGraw-Hill, 630 pp.

Durbin, T. J. 1978. *Calibration of a Mathematical Model of the Antelope Valley Ground-Water Basin, California*. (U.S. Geol. Survey Water Supply Paper 2046.) 51 pp.

Faust, C. R., and J. W. Mercer. 1980a. "Ground-Water Modeling: Numerical Models." *Ground Water* 18(4):395–409.

Faust, C. R., and J. W. Mercer. 1980b. "Ground-Water Modeling: Recent Developments." *Ground Water* 18(6):569–577.

Fetter, C. W. 1980. *Applied Hydrogeology*. Columbus, Ohio: Merrill, 488 pp.

Finnemore, E. J., and B. Perry. 1968. "Seepage Through an Earth Dam Computed by the Relaxation Technique." *Water Resources Research* 4(5):1059–1067.

Fleck, W. B., and M. G. McDonald. 1978. "Three-Dimensional Finite-Difference Model of Ground-Water System Underlying the Muskegon County Wastewater Disposal System, Michigan." *U.S. Geol. Survey Journal of Research* 6(3):307–318.

Forsythe, G. E., and W. Wasow. 1960. *Finite Difference Methods for Partial Differential Equations.* New York: Wiley, 444 pp.

Freeze, R. A. 1971. "Three-Dimensional, Transient, Saturated-Unsaturated Flow in a Groundwater Basin." *Water Resources Research* 7(2):347–366.

Freeze, R. A., and J. A. Cherry. 1979. *Groundwater.* Englewood Cliffs, N.J.: Prentice-Hall, 604 pp.

Freeze, R. A., and P. A. Witherspoon. 1966. "Theoretical Analysis of Regional Ground-water Flow, 1: Analytical and Numerical Solutions to the Mathematical Model." *Water Resources Research* 2(4):641–656.

Freeze, R. A., and P. A. Witherspoon. 1967. "Theoretical Analysis of Regional Ground-water Flow, 2: Effect of Water-Table Configuration and Subsurface Permeability Variation." *Water Resources Research* 3(2):623–634.

Freeze, R. A., and P. A. Witherspoon. 1968. "Theoretical Analysis of Regional Groundwater Flow, 3: Quantitative Interpretations." *Water Resources Research* 4(3):581–590.

Fried, J. J. 1975. *Groundwater Pollution.* Amsterdam: Elsevier Scientific, 330 pp.

Gelhar, L. W., A. L. Gutjahr, and R. L. Naff. 1979. "Stochastic Analysis of Macrodispersion in a Stratified Aquifer." *Water Resources Research* 15(6):1387–1397.

Getzen, R. T. 1977. *Analog-Model Analysis of Regional Three-Dimensional Flow in the Ground-Water Reservoir of Long Island, New York.* (U.S. Geol. Survey Professional Paper 982.) 49 pp.

Gillham, R. W., and R. N. Farvolden. 1974. "Sensitivity Analysis of Input Parameters in Numerical Modeling of Steady State Regional Groundwater Flow." *Water Resources Research* 10(3):529–538.

Gray, W. G. 1976. "The Finite Element Method in Groundwater Transport." In *Advances in Groundwater Hydrology*, Z. A. Saleem, ed. Minneapolis: American Water Resources Association, 130–143.

Gray, W. G., and G. F. Pinder. 1974. "Galerkin Approximation of the Time Derivative in the Finite Element Analysis of Groundwater Flow." *Water Resources Research* 10(4):821–828.

Gray, W. G., and G. F. Pinder. 1976a. "On the Relationship Between the Finite Element and Finite-Difference Methods." *International Journal of Numerical Methods in Engineering* 10:893–923.

Gray, W. G., and G. F. Pinder. 1976b. "An Analysis of the Numerical Solution of the Transport Equation." *Water Resources Research* 12(3):547–555.

Grove, D. B. 1976. "Ion Exchange Reactions Important in Groundwater Quality Models." In *Advances in Groundwater Hydrology*, Z. A. Saleem, ed. Minneapolis: American Water Resources Association, 144–152.

Grove, D. B. 1977. *The Use of Galerkin Finite-Element Methods to Solve Mass-Transport Equations.* (U.S. Geol. Survey Water Resources Investigations, 77–49.) 61 pp.

Gureghian, A. B., D. S. Ward, and R. W. Cleary. 1981. "A Finite Element Model for the Migration of Leachate from a Sanitary Landfill in Long Island, New York—Part II: Application." *Water Resources Bulletin* 17(1):62–66.

Gutjahr, A. L., and L. W. Gelhar. 1981. "Stochastic Models of Subsurface Flow: Infinite Versus Finite Domains and Stationarity." *Water Resources Research* 17(2):337–350.

Gutjahr, A. L., et al. 1978. "Stochastic Analysis of Spatial Variability in Subsurface Flows, 2: Evaluation and Application." *Water Resources Research* 14(5):953–959.

Guymon, G. L. 1970. "A Finite Element Solution of the One-Dimensional Diffusion-Convection Equation." *Water Resources Research* 6(1):204–210.

Hodge, R. A. L., and R. A. Freeze. 1977. "Groundwater Flow Systems and Slope Stability." *Canadian Geotechnical Journal* 14:466–476.

Hubbert, M. K. 1940. "The Theory of Groundwater Motion." *Journal of Geology* 48: 785–944.

Hubbert, M. K. 1969. *The Theory of Ground-Water Motion and Related Papers.* New York: Hafner, 310 pp.

Huebner, K. H. 1975. *The Finite Element Method for Engineers.* New York: Wiley, 500 pp.

Huntoon, P. W. 1974. *Finite Difference Methods As Applied to the Solution of Groundwater Flow Problems.* Laramie: Wyoming Water Resources Research Institute, University of Wyoming, 108 pp.

Ingersoll, L. R., O. J. Zobel, and A. C. Ingersoll. 1954. *Heat Conduction with Engineering, Geological and Other Applications* (Revised Edition). Madison: University of Wisconsin Press, 325 pp.

Ivanovich, M., and D. B. Smith. 1978. "Determination of Aquifer Parameters by a Two Well Pulsed Method Using Radioactive Tracers." *Journal of Hydrology* 36:35–45.

Jacob, C. E. 1943. "Correlation of Ground-Water Levels and Precipitation on Long Island, New York." *Trans. Amer. Geophysical Union,* 564–573.

Karanjac, J., M. Altunkaynak, and G. Ovul. 1977. "Mathematical Model of Uluova Plain, Turkey—A Training and Management Tool." *Ground Water* 15(5):348–357.

Kirkham, D. 1967. "Explanation of Paradoxes in Dupuit–Forchheimer Seepage Theory." *Water Resources Research* 3(2):609–622.

Konikow, L. F. 1977. *Modeling Chloride Movement in the Alluvial Aquifer at the Rocky Mountain Arsenal, Colorado.* (U.S. Geol. Survey, Water Supply Paper 2044.) 43 pp.

Konikow, L. F., and J. D. Bredehoeft. 1974. "Modeling Flow and Chemical Quality Changes in an Irrigated Stream-Aquifer System." *Water Resources Research* 10(3):546–562.

Konikow, L. F., and J. D. Bredehoeft. 1978. *Computer Model of Two-Dimensional Solute Transport and Dispersion in Ground Water.* (U.S. Geol. Survey, Techniques of Water Resources Investigations Book 7, Chapter C2.) 90 pp.

Konikow, L. F., and D. B. Grove. 1977. *Derivation of Equations Describing Solute Transport in Ground Water.* (U.S. Geol. Survey, Water Resources Investigations, 77-19.) 30 pp.

Land, L. F. 1977. "Utilizing a Digital Model to Determine the Hydraulic Properties of a Layered Aquifer." *Ground Water* 15(2):153–159.

Larson, S. P. 1978. *Direct Solution Algorithm for the Two-Dimensional Ground-Water Flow Model.* (U.S. Geol. Survey, Open-File Report 79-202.) 28 pp.

Leake, S. A. 1977. "Simulation of Flow from an Aquifer to a Partially Penetrating Trench." *U.S. Geol. Survey Journal of Research* 5(5):535–540.

Lehr, J. H. 1979. "Editorial: Mathematical Ground-Water Models May Be Intellectual Toys Today, but They Should Be Useful Tools Tomorrow." *Ground Water* 17(5):418–422.

McBride, M. S., and H. O. Pfannkuch. 1975. "The Distribution of Seepage Within Lake-beds." *U.S. Geol. Survey Journal of Research* 3(5):505–512.

Matheron, G., and G. De Marsily. 1980. "Is Transport in Porous Media Always Diffusive?" *Water Resources Research* 16(5):901–917.

Mercer, J. W., and C. R. Faust. 1980a. "Ground-Water Modeling: An Overview." *Ground Water* 18(2):108–115.

Mercer, J. W., and C. R. Faust. 1980b. "Ground-Water Modeling: Mathematical Models." *Ground Water* 18(3):212–227.

Mercer, J. W., and C. R. Faust. 1980c. "Ground-Water Modeling: Applications." *Ground Water* 18(5):486–497.

Mercer, J. W., and C. R. Faust. 1981. *Ground-Water Modeling.* Worthington, Ohio: National Water Well Association, 60 pp.

Mercer, J. W., G. F. Pinder, and I. G. Donaldson. 1975. "A Galerkin Finite Element Analysis of the Hydro-Thermal System at Wairakie, New Zealand." *Journal of Geophysical Research* 80(17):2608–2621.

Murray, W. A., and P. L. Monkmeyer. 1973. "Validity of Dupuit–Forchheimer Equation." *Journal of the Hydraulics Division* (Amer. Society of Civil Engineers) 97(HY9):1573–1583.

Neuman, S. P. 1973. "Calibration of Distributed Parameter Groundwater Flow Models Viewed as a Multiple-Objective Decision Process Under Uncertainty." *Water Resources Research* 9(4):1006–1021.

Neuman, S. P. 1976. *User's Guide for FREESURF I.* Tucson: Dept. of Hydrology and Water Resources, University of Arizona, 22 pp.

Neuman, S. P. 1980. "A Statistical Approach to the Inverse Problem of Aquifer Hydrology, 3: Improved Solution Method and Added Perspective." *Water Resources Research* 16(2):331–346.

Neuman, S. P., and P. A. Witherspoon. 1970. "Finite Element Method of Analyzing Steady Seepage with a Free Surface." *Water Resources Research* 6(3):889–897.

Neuman, S. P., and P. A. Witherspoon. 1971. "Analysis of Nonsteady Flow with a Free Surface Using the Finite Element Method." *Water Resources Research* 7(3):611–623.

Neuman, S. P., and S. Yakowitz. 1979. "A Statistical Approach to the Inverse Problem of Aquifer Hydrology, 1: Theory." *Water Resources Research* 15(4):845–860.

Neuman, S. P., G. E. Fogg, and E. A. Jacobson. 1980. "A Statistical Approach to the Inverse Problem of Aquifer Hydrology, 2: Case Study." *Water Resources Research* 16(1):33–58.

Ogata, A., and R. B. Banks. 1961. *A Solution of the Differential Equation of Longitudinal Dispersion in Porous Media.* (U.S. Geol. Survey Professional Paper 411-A.) 7 pp.

Pickens, J. F., and W. C. Lennox. 1976. "Numerical Simulation of Waste Movement in Steady Groundwater Flow Systems." *Water Resources Research* 12(2):171–180.

Pickens, J. F., W. F. Merritt, and J. A. Cherry. 1980. "Field Determination of the Physical Contaminant Transport Parameters in a Sandy Aquifer." In *Nuclear Techniques in Groundwater Pollution Research.* Vienna: International Atomic Energy Agency, 239–265.

Pinder, G. F. 1973. "A Galerkin Finite Element Simulation of Groundwater Contamination on Long Island, N.Y." *Water Resources Research* 9(6):1657–1669.

Pinder, G. F., and H. H. Cooper, Jr. 1970. "A Numerical Technique for Calculating the Transient Position of the Saltwater Front." *Water Resources Research* 6(3):875–882.

Pinder, G. F., and E. O. Frind. 1972. "Application of Galerkin's Procedure to Aquifer Analysis." *Water Resources Research* 8(1):108–120.

Pinder, G. F., E. O. Frind, and S. S. Papadopulos. 1973. "Functional Coefficients in the Analysis of Groundwater Flow." *Water Resources Research* 9(1):222–226.

Pinder, G. F., and W. G. Gray. 1976. "Is There a Difference in the Finite Element Method?" *Water Resources Research* 12(1):105–107.

Pinder, G. F., and W. G. Gray. 1977. *Finite Element Simulation in Surface and Subsurface Hydrology*. New York: Academic Press, 295 pp.

Plummer, L. N. 1977. "Defining Reactions and Mass Transfer in Part of the Floridan Aquifer." *Water Resources Research* 13(5):801–812.

Prickett, T. A. 1975. "Modeling Techniques for Groundwater Evaluation." In *Advances in Hydroscience*, Vol. 10. New York: Academic Press, 1–143.

Prickett, T. A., and C. G. Lonnquist. 1971. *Selected Digital Computer Techniques for Groundwater Resource Evaluation*. (Illinois State Water Survey Bulletin 55.) 62 pp.

Remson, I., G. M. Hornberger, and F. J. Molz. 1971. *Numerical Methods in Subsurface Hydrology*. New York: Wiley-Interscience, 389 pp.

Robson, S. G. 1978. *Application of Digital Profile Modeling Techniques to Ground-Water Solute Transport at Barstow, California*. (U.S. Geol. Survey Water Supply Paper 2050.) 28 pp.

Rubin, J., and R. V. James. 1973. "Dispersion-Affected Transport of Reacting Solutes in Saturated Porous Media: Galerkin Method Applied to Equilibrium-Controlled Exchange in Unidirectional Steady Water Flow." *Water Resources Research* 9(5):1332–1356.

Rushton, K. R., and S. C. Redshaw. 1979. *Seepage and Groundwater Flow*. New York: Wiley, 339 pp.

Rushton, K. R., and L. M. Tomlinson. 1977. "Permissible Mesh Spacing in Aquifer Problems Solved by Finite Differences." *Journal of Hydrology* 34:63–76.

Schwartz, F. W. 1975. "On Radioactive Waste Management: An Analysis of the Parameters Controlling Subsurface Contaminant Transfer." *Journal of Hydrology* 27:51–71.

Schwartz, F. W. 1977. "On Radioactive Waste Management: Model Analysis of a Proposed Site." *Journal of Hydrology* 32:257–277.

Segerlind, L. J. 1976. *Applied Finite Element Analysis*. New York: Wiley, 422 pp.

Segol, G., and G. F. Pinder. 1976. "Transient Simulation of Salt Water Intrusion in Southeastern Florida." *Water Resources Research* 12(1):65–70.

Segol, G., G. F. Pinder, and W. G. Gray. 1975. "A Galerkin Finite Element Technique for Calculating the Transient Position of the Saltwater Front." *Water Resources Research* 11(2):343–347.

Smith, G. D. 1965. *Numerical Solution of Partial Differential Equations*. New York: Oxford University Press, 179 pp.

Smith, L., and R. A. Freeze. 1979. "Stochastic Analysis of Steady State Groundwater Flow in a Bounded Domain, 2: Two-Dimensional Simulations." *Water Resources Research* 15(6):1543–1559.

Smith, L., and F. W. Schwartz. 1980. "Mass Transport, 1: A Stochastic Analysis of Macroscopic Dispersion." *Water Resources Research* 16(2):303–313.

Smith, L., and F. W. Schwartz. 1981. "Mass Transport, 2, Analysis of Uncertainty in Prediction." *Water Resources Research* 17(2):351–369.

Stone, H. L. 1968. "Interactive Solution of Implicit Approximations of Multidimensional Partial Differential Equations." *SIAM Journal of Numerical Analysis* 5:530–558.

Tang, D. H., and G. F. Pinder. 1979. "Analysis of Mass Transport with Uncertain Physical Parameters." *Water Resources Research* 15(5):1147–1155.

Theis, C. V. 1935. "The Relation Between the Lowering of the Piezometric Surface and the

Rate and Duration of Discharge of a Well Using Groundwater Storage." *Trans. Amer. Geophys. Union* 2:519–524.

Toth, J. 1962. "A Theory of Groundwater Motion in Small Drainage Basins in Central Alberta, Canada." *Journal of Geophysical Research* 67(11):4375–4387.

Toth, J. 1963. "A Theoretical Analysis of Groundwater Flow in Small Drainage Basins." *Journal of Geophysical Research* 68(16):4795–4812.

Trescott, P. C. 1975. *Documentation of Finite-Difference Model for Simulation of Three-Dimensional Ground-Water Flow.* (U.S. Geol. Survey Open File Report 75-438.) 32 pp.

Trescott, P. C., and S. P. Larson. 1976. *Supplement to Open File Report 75-438.* (U.S. Geol. Survey Open File Report 76-591.)

Trescott, P. C., and S. P. Larson. 1977. "Solution to Three-Dimensional Groundwater Flow Equations Using the Strongly Implicit Procedure." *Journal of Hydrology* 35:49–60.

Trescott, P. C., G. F. Pinder, and S. P. Larson. 1976. *Finite-Difference Model for Aquifer Simulation in Two Dimensions with Results of Numerical Experiments.* (U.S. Geol. Survey Techniques of Water Resources Investigations, Book 7, Chapter C1.) 116 pp.

Turk, G. 1978. Discussion of a paper by Karanjac et al. (1977) with reply by Karanjac et al. *Ground Water* 16(3):207–210.

Verma, R. D., and W. Brutsaert. 1971. "Unsteady Free Surface Ground Water Seepage." *Journal of the Hydraulics Division* (American Society of Civil Engineers) 97(HY8): 1213–1229.

Wang, H. F., and M. P. Anderson. 1977. "Finite Differences and Finite Elements as Weighted Residual Solutions to Laplace's Equation." In *Finite Elements in Water Resources,* W. G. Gray, G. F. Pinder, and C. A. Brebbia, eds. London: Pentech Press, 2.167–2.178.

Webster, D. S., J. F. Proctor, and I. W. Marine. 1970. *Two-Well Tracer Test in Fractured Crystalline Rock.* (U.S. Geol. Survey Water Supply Paper 1544-I.) 22 pp.

Winter, T. C. 1976. *Numerical Simulation Analysis of the Interaction of Lakes and Groundwater.* (U.S. Geol. Survey Professional Paper 1001.) 45 pp.

Winter, T. C. 1978. "Numerical Simulation of Steady State Three-Dimensional Groundwater Flow near Lakes." *Water Resources Research* 14(2):245–254.

Zienkiewicz, O. C. 1977. *The Finite Element Method* (3rd ed.). New York: McGraw-Hill, 787 pp.

Zienkiewicz, O. C., and Y. K. Cheung. 1965. "Finite Elements in the Solution of Field Problems." *The Engineer* 220(5722):507–510.

Index